Lecture Notes in Applied and Computational Mechanics

Volume 53

Series Editors

Prof. Dr.-Ing. Friedrich Pfeiffer
Prof. Dr.-Ing. Peter Wriggers

Lecture Notes in Applied and Computational Mechanics

Edited by F. Pfeiffer and P. Wriggers

Further volumes of this series found on our homepage: springer.com

Vol. 53: Litewka, P.
Finite Element Analysis of Beam-to-Beam Contact
159 p. 2010 [978-3-642-12939-1]

Vol. 52: Pilipchuk, V.N.
Nonlinear Dynamics: Between Linear and Impact Limits
364 p. 2010 [978-3-642-12798-4]

Vol. 51: Besdo, D., Heimann, B., Klüppel, M., Kröger, M., Wriggers, P., Nackenhorst, U.
Elastomere Friction
249 p. 2010 [978-3-642-10656-9]

Vol. 50: Ganghoffer, J.-F., Pastrone, F. (Eds.)
Mechanics of Microstructured Solids 2
102 p. 2010 [978-3-642-05170-8]

Vol. 49: Hazra, S.B.
Large-Scale PDE-Constrained Optimization in Applications
224 p. 2010 [978-3-642-01501-4]

Vol. 48: Su, Z.; Ye, L.
Identification of Damage Using Lamb Waves
346 p. 2009 [978-1-84882-783-7]

Vol. 47: Studer, C.
Numerics of Unilateral Contacts and Friction
191 p. 2009 [978-3-642-01099-6]

Vol. 46: Ganghoffer, J.-F., Pastrone, F. (Eds.)
Mechanics of Microstructured Solids
136 p. 2009 [978-3-642-00910-5]

Vol. 45: Shevchuk, I.V.
Convective Heat and Mass Transfer in Rotating Disk Systems
300 p. 2009 [978-3-642-00717-0]

Vol. 44: Ibrahim R.A., Babitsky, V.I., Okuma, M. (Eds.)
Vibro-Impact Dynamics of Ocean Systems and Related Problems
280 p. 2009 [978-3-642-00628-9]

Vol.43: Ibrahim, R.A.
Vibro-Impact Dynamics
312 p. 2009 [978-3-642-00274-8]

Vol. 42: Hashiguchi, K.
Elastoplasticity Theory
432 p. 2009 [978-3-642-00272-4]

Vol. 41: Browand, F., Ross, J., McCallen, R. (Eds.)
Aerodynamics of Heavy Vehicles II: Trucks, Buses, and Trains
486 p. 2009 [978-3-540-85069-4]

Vol. 40: Pfeiffer, F.
Mechanical System Dynamics
578 p. 2008 [978-3-540-79435-6]

Vol. 39: Lucchesi, M., Padovani, C., Pasquinelli, G., Zani, N.
Masonry Constructions: Mechanical Models and Numerical Applications
176 p. 2008 [978-3-540-79110-2]

Vol. 38: Marynowski, K.
Dynamics of the Axially Moving Orthotropic Web
140 p. 2008 [978-3-540-78988-8]

Vol. 37: Chaudhary, H., Saha, S.K.
Dynamics and Balancing of Multibody Systems
200 p. 2008 [978-3-540-78178-3]

Vol. 36: Leine, R.I.; van de Wouw, N.
Stability and Convergence of Mechanical Systems with Unilateral Constraints
250 p. 2008 [978-3-540-76974-3]

Vol. 35: Acary, V.; Brogliato, B.
Numerical Methods for Nonsmooth Dynamical Systems: Applications in Mechanics and Electronics
545 p. 2008 [978-3-540-75391-9]

Vol. 34: Flores, P.; Ambrósio, J.; Pimenta Claro, J.C.; Lankarani Hamid M.
Kinematics and Dynamics of Multibody Systems with Imperfect Joints: Models and Case Studies
186 p. 2008 [978-3-540-74359-0

Vol. 33: Nies ony, A.; Macha, E.
Spectral Method in Multiaxial Random Fatigue
146 p. 2007 [978-3-540-73822-0]

Vol. 32: Bardzokas, D.I.; Filshtinsky, M.L.; Filshtinsky, L.A. (Eds.)
Mathematical Methods in Electro-Magneto-Elasticity
530 p. 2007 [978-3-540-71030-1]

Vol. 31: Lehmann, L. (Ed.)
Wave Propagation in Infinite Domains
186 p. 2007 [978-3-540-71108-7]

Vol. 30: Stupkiewicz, S. (Ed.)
Micromechanics of Contact and Interphase Layers
206 p. 2006 [978-3-540-49716-5]

Finite Element Analysis of Beam-to-Beam Contact

Przemysław Litewka

Springer

Przemysław Litewka
Institute of Structural Engineering
Poznan University of Technology
ul. Piotrowo 5
60-965 Poznan
Poland
E-mail: przemyslaw.litewka@put.poznan.pl

ISBN: 978-3-642-12939-1 e-ISBN: 978-3-642-12940-7

DOI 10.1007/ 978-3-642-12940-7

Lecture Notes in Applied and Computational Mechanics ISSN 1613-7736

e-ISSN 1860-0816

Library of Congress Control Number: 2010926492

Typeset & Cover Design: Scientific Publishing Services Pvt. Ltd., Chennai, India.

Printed on acid-free paper

9 8 7 6 5 4 3 2 1 0

springer.com

Abstract

This work concerns a multi-aspect finite element analysis of beam-to-beam contact. Specific features of this type of contact are discussed and compared to the case of contact between typical 2D and 3D solids. Beams with rectangular and circular cross-sections are considered. The Coulomb law of friction in a framework of the analogy to plasticity is taken into account. A formulation of smooth contact elements using the Hermite's polynomials and the Bezier's curves is presented. Some aspects of a coupling between fields of displacements, temperature and electric voltage are considered, too. Several types of beam-to-beam contact elements are derived. Their performance is checked solving numerous examples using self-written computer programs. The novelty achievements in this work include: formulation of contact search and check routine for beams with rectangular cross-sections, derivation of four types of versatile smooth beam-to-beam contact finite elements and a preliminary analysis of a thermo-electro-mechanical coupling for the beam-to-beam contact.

Przemysław Litewka

Preface

Phenomena occurring during a contact of two bodies are encountered in everyday life. In reality almost every type of motion is related to frictional contact between a moving body and a ground. Moreover, modeling of simple and more complex processes as nailing, cutting, vacuum pressing, movement of machines and their elements, rolling or, finally, a numerical simulation of car crash tests, requires taking contact into account. Therefore, its analysis has been a subject of many research efforts for a long time now. However, only recent developments of computer power and the related progress in numerical methods, especially the finite element method, allowed to treat contact problems in a more accurate way. To this end a precise mathematical description of surface of contacting bodies, advanced physical models of materials involved and of phenomena in the contact points as well as suitable numerical methods to solve problems with large number of unknowns are necessary.

Including contact itself introduces a non-linearity to a system related to a change of static scheme. Besides, in many cases it is necessary to analyze inelastic bodies undergoing large displacements and (or) large strains. Then the problem in hand becomes strongly non-linear and can be solved effectively only using numerical methods.

Another difficulty is related to a definition of a physical model of phenomena occurring at contact spots if heat conduction and electric current flow are involved. In this case coupling of mechanical, thermal and electric fields plays an important role and the model becomes even more complicated.

This outline points out the wide range of problems that are involved in the contact modeling. In the recent years a lot of scientific contributions were published concerning those aspects of the problem. However, it is author's opinion that there are relatively few efforts related to contact between structural elements, like beams, plates or shells. The purpose of this work is to fill this gap. It concerns the beam-to-beam contact as a specific case of the 3D solids contact. A numerical formulation of frictional contact for beams with two shapes of cross-section is derived. Further, a couple of effective methods for modeling of smooth curves representing beam axes are presented. A part of the book is also devoted to analyze some aspects of thermo-electro-mechanical coupling in contact of thermal and electric conductors. Analyses in every chapter are illustrated with numerical examples showing the performance of derived contact finite elements.

This book is a summary of a research spanning recent few years. Its beginning is related to my scientific stay at the University of Hanover. I would like to thank Prof. Jerzy Rakowski from my institute for the incentive and help to organize it, Prof. Peter Wriggers from Hanover for the hospitality and scientific help he kindly offered me in his institute as well as my colleagues and co-workers from Hanover and Padua: Dr. Daniela Boso, Dr. Lovre Krstulović-Opara and Prof. Bernhard Schrefler, for a fruitful friendly collaboration.

Finally I wish to express my thanks to my family, my beloved Wife and dear Parents for love and unceasing motivation to work.

Przemysław Litewka

Contents

Chapter 1
Introduction

1.1 From the Ancient Egypt to the Computer Era

Contact and related phenomena occurring at the interface of two bodies are encountered at every step of everyday life, in nature and in engineering. They are so common, that usually one does not think about them, despite the fact one takes advantage of their effects. For instance, such a common activity as walking does require frictional contact between shoes and a floor. Similarly, without friction no vehicle would move, no nail would hold to a wall, all objects on even little inclined surfaces would slide down.

Friction is a phenomenon strictly connected to contact. In the situations mentioned above it is desirable but in other situations it may be opposite and we may wish to decrease it. One of the oldest methods to achieve this end can be seen on Egyptian paintings showing transport of rock blocks used to construct monumental objects. There is a person depicted, who pours a liquid, probably water, in front of sledges used to transport this construction material.

First scientific experiments with contact are due to Leonardo da Vinci who concluded, that the friction force F_T depends on the body weight F_N and does not depend on the contact area. Later it was put by Euler (1748a, 1748b) in a formula

$$F_T = \mu F_N \,, \tag{1.1}$$

which is currently known as the Coulomb friction law. Coulomb (1779) carried out further experiments and introduced adhesion to get the enhanced formula

$$F_T = A + \mu F_N \,. \tag{1.2}$$

Development of theory of elasticity at the break of 19th and 20th centuries led to research by Hertz (1882), who managed to solve analytically several cases of elastic bodies in contact. These solutions can be found in monographs by Johnson (1985) or Kikuchi and Oden (1988).

However, analytical methods used in the contact analysis proved to be of little use due to the non-linearity and complicated boundary conditions. Only the development of computers and the finite element method starting in 1950s provided the scientists with a tool suitable to treat contact problems. First efforts concerned a linear theory (Wilson and Parsons 1970, Chan and Tuba 1971) but the quick

P. Litewka: Finite Element Analysis of Beam-to-Beam Contact, LNACM 53, pp. 1–11.
springerlink.com

progress of numerical methods and computer technique followed. This enabled effective treatment of complicated phenomena in a contact zone including coupling of mechanical, thermal and electric fields in the non-linear range. First papers dealing with application of the finite element method in the contact analysis with large strains involved were written by: Curnier and Alart (1988), Simo and Laursen (1992) and Wriggers and Miehe (1992). Among many other known publications it is worth to mention two, almost simultaneously published, monographs by Laursen (2002) and by Wriggers (2002). They include a thorough presentation of mathematical background of numerical contact formulation, physical models of phenomena at the contact interfaces and the theoretical considerations are illustrated by numerical results concerning many practical cases of contacting bodies.

Besides the finite element method there exist also some attempts to use the boundary element method in the contact analysis. However, due to the limitations of the method, they involve only cases with small displacements and small strains (Gun 2004).

An intensive development of meshless methods observed in recent years also led to some applications in the field of contact (Li et al. 2001).

Another approach, where a combination of numerical and analytical methods was tested, can be found in the work by Li and Berger (2003) who analyzed the frictional contact of 3D solids.

Contact is also a field of a pure mathematical analysis. There is a group of authors who deal with such important aspects as existence and uniqueness of solutions. Among many important contributions to this area are those by: Moreau (1974), Klarbring (1988, 1990) or Pinto da Costa and Martins (2003). It was also found, that contact can be treated by the mathematical programming algorithms (Klarbring 1986, Klarbring and Björkman 1988, Björkman et al. 1995, Kim and Kwak 1996, Sung and Kwak 2002). These methods were also used in the analysis of so called slackened bar systems (Gawęcki 1987, Gawęcki and Janińska 1995, Gawęcki and Kuczma 1995 as well as Gawęcki et al. 1998). A detailed description of these methods, which were originally devised in the optimization theory, can be found for instance in the monograph by Luenberger (1984).

Yet another aspect of numerical contact analysis concerns a mathematical description of geometry of contacting facets. These problems are thoroughly described in the monograph by Farin (1993).

A vast range of research results and scientific publications does not cover to much extent the cases of contact between beams. The few papers include the one by Maker and Laursen (1994), who analyzed contact between a beam and a rigid body, and those by Wriggers and Zavarise (1997) and Zavarise and Wriggers (2000), which were devoted to frictionless and frictional contact of beams with circular cross-sections. The reason for the scarcity of this research is definitely a much smaller number of practical applications than in the case of typical solids. Yet, the situations exist, where application of simpler analysis including beam contact finite elements can be profitable. One might think of contact between moving elements of machines, woven fabrics and geotextiles, sport racquet stringing or, finally, a multi-degree twists of strands in steel cables or electric conductors. Especially the last case of beam-to-beam contact poses a great challenge because it requires an advanced numerical model including effects of coupling of displacements, temperature, electric current and magnetic field.

Beam contact is a specific case and is characterized by some features, which demand a different approach than in the case of typical solids. One of the purposes of this book is to present these differences.

1.2 Frictionless Contact between Solids

Let us consider two bodies, B_1 and B_2, shown in Fig. 1.1, which undergo deformation *f*, such that any arbitrary points on their surfaces, described in the initial configuration with position vectors $\mathbf{X}_1$ and $\mathbf{X}_2$ undergo displacements $\mathbf{u}_1$ and $\mathbf{u}_2$. In the current configuration they are defined by the position vectors $\mathbf{x}_1$ and $\mathbf{x}_2$, what can be expressed by the formulae

$$f:\begin{cases}\mathbf{x}_1 = \mathbf{X}_1 + \mathbf{u}_1 \\ \mathbf{x}_2 = \mathbf{X}_2 + \mathbf{u}_2\end{cases}. \tag{1.3}$$

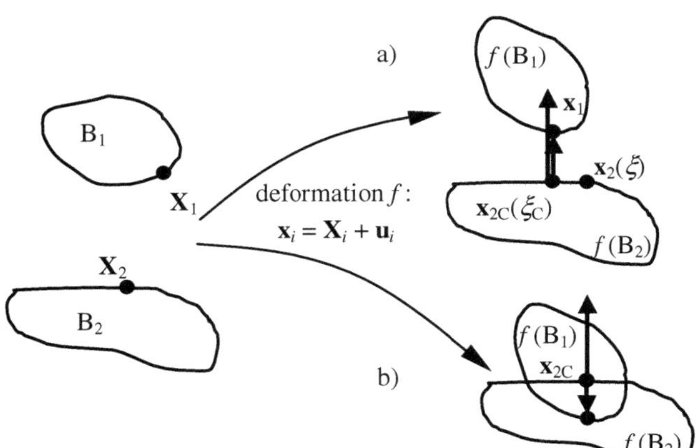

Fig. 1.1 Deformation of two bodies and penetration function: a) separate bodies, b) penetration.

Contact concerns exclusively points on surface of bodies and its formulation is practically independent of a physical model of the body material. That is why the phenomena occurring inside the bodies are not dealt with here. It is only assumed that the deformation *f* leads to large displacements and large strains in both bodies.

Solution of a problem in the theory of elasticity concerning the two bodies involves finding a minimum of the potential energy functional Π

$$\min \Pi = \min(\Pi_1 + \Pi_2). \tag{1.4}$$

Theoretically, in the deformation process *f*, a situation presented in Fig. 1.1b may occur, when the bodies intersect mutually. Obviously, in reality such a state is not allowed. To exclude such a possibility, unilateral constraints are introduced. They are formulated using a penetration function (or a gap function) g_N. To this end one of the bodies, e.g. B_1, is distinguished as a slave body. Then the points $\mathbf{x}_1$ are projected on the surface of the second body, i.e. B_2, which is called a master body. In this way pairs of points $\mathbf{x}_1$ and $\mathbf{x}_{2C}$ are obtained, as in Fig. 1.1. The location of the point $\mathbf{x}_{2C}$ is expressed in terms of local surface co-ordinates $\xi = \xi_C$. Then the penetration function is defined as

$$g_N = \mathbf{n} \circ (\mathbf{x}_1 - \mathbf{x}_{2C}), \tag{1.5}$$

where the unit outward normal vector $\mathbf{n}$ on the surface of the body B_2 is introduced (Fig. 1.1) and „$\circ$" denotes the scalar product of vectors. The value of the function g_N provides the distance between the points and in the case of penetration – its depth.

Preserving the condition of non-penetrability requires that the penetration function remains non-negative

$$g_N \geq 0. \tag{1.6}$$

The condition expressed by Eq. (1.6) constitutes an inequality constraint, which must be included in the minimization of the functional (1.4). It should be emphasized, that the inequality (1.6) concerns all possible point pairs $\mathbf{x}_1$–$\mathbf{x}_{2C}$.

If the condition (1.6) is accompanied by the constraint of contact normal force, which can only be compressive

$$F_N \leq 0, \tag{1.7}$$

then the complete set of the classical Hertz-Signorini-Moreau conditions for the frictionless contact is obtained, together with the equality

$$F_N g_N = 0. \tag{1.8}$$

According to (1.7), in the case of contact the penetration is zero and the contact force is negative, while in the case of separation the gap is positive and the contact force is zero (Moreau 1974). These are the conditions of unilateral contact (Johnson 1985, Kikuchi and Oden 1988), called also the Kuhn-Tucker conditions in the theory of optimization.

The relations yielding from contact, presented above, introduce only geometric constraints. Such a case of contact may be called a low precision contact. In some situations it is necessary to consider high precision contact, which requires introduction of physical law for the microscale phenomena at rough contacting surfaces. These laws usually include a relation between the loading of the contact area and the mean distance between the bodies. Such a relation can be obtained experimentally (Greenwood and Williamson 1966, Kragelsky et al. 1982) or using a statistic analysis (Song and Yovanovich 1987). The latter approach was also used by Zavarise (1991) to find an exponential form of this relation. In this book considerations are limited to the low precision contact.

1.3 Methods of Introduction of Contact Constraints

The problem of the low precision contact leads to a solution of a functional minimization with inequality constraints. The overview of mathematical methods suitable to solve this minimization can be found, for instance, in the monographs on computational contact (Wriggers 2002, Laursen 2002), while the thorough description of methods based on linear and non-linear programming is given in the book by Luenberger (1984). Here only three of these methods are considered.

The starting point for all of them is a concept of an active set. Among all possible pairs $\mathbf{x}_1$–$\mathbf{x}_{2C}$, for which the penetration function g_N is defined, only those fulfilling the condition $g_N < 0$ are selected. In a computational approach a contact search procedure is applied, which, on the base of geometric considerations, finds the contact spots, where the inequality condition for separation (1.6) is not fulfilled. For these pairs $\mathbf{x}_1$–$\mathbf{x}_{2C}$, which compose the active set the inequality constraints can be replaced by the equality constraints

$$g_N = 0 \tag{1.9}$$

and the remaining pairs are not considered at all. It should be pointed out, that the incremental-iterative procedures used in the computational contact analysis, e.g. the Newton-Raphson method, require updates of the active set at each iteration.

In this way the problem in hand takes a form of a functional minimization with equality constraints. Such a problem is known from the theory of optimization and can be solved by the well known methods – the penalty method, the Lagrange multipliers method or the augmented Lagrangian method. These three methods are briefly discussed here.

The penalty method leads to the following modification of the functional (1.4) subjected to the minimization

$$\min\left(\Pi + \sum_{\text{act}} \frac{1}{2} \varepsilon_N g_N{}^2\right). \tag{1.10}$$

The sum in this formula concerns all active contacting pairs of points. The value of the coefficient ε_N is usually determined by the "trial-and-error" approach for the problem in hand. It is so, because the constraint condition (1.9) is fulfilled only, when $\varepsilon_N = \infty$. In practice the penalty coefficient has a large but finite value and it is accepted that

$$|g_N| = e < e_{\text{tol}}, \tag{1.11}$$

where e_{tol} is a certain agreed tolerance determining an acceptable value of penetration.

However, it was proved, that ε_N cannot take arbitrarily large values because this leads to a numerical instability of the algorithm, what is manifested in a divergence of iterative solution procedures. That is why the application of the penalty method requires a compromise between the necessity to fulfil the condition (1.11) and the possibility to solve the problem.

It can be noted that the penalty parameter has a physical interpretation. It represents the stiffness of a fictitious spring, which joins the points of an active pair, $\mathbf{x}_1$ and $\mathbf{x}_{2C}$. Because this constraint is unilateral this spring can only be compressed (1.7). In order to fulfil the constraint condition (1.9), this spring would need to be of infinite stiffness. This interpretation leaves some space for an interesting approach, where the stiffness value might be adopted in a way to represent real elastic properties of surfaces of contacting bodies.

This interpretation allows to identify the source of convergence problems of iterative solution schemes, if the penalty parameter has too large value. If this value is adopted correctly the force in the fictitious spring is able to reduce the penetration in the following iteration to a value close to zero. On the other hand, if the stiffness is too large, then the force in the spring becomes so big, that the correction is too large, what may lead to a separation of the bodies in the following iteration. Then the contact search procedure excludes the given pair from the active set. Next, the external loads bring the pair back to contact and the infinite cycle starts, impeding the convergence to the equilibrium state.

The second, very frequently used method to include the equality constraints is the Lagrange multiplier method. In this approach the saddle-point problem of the modified functional is encountered and this leads to the stationarity point formulation

$$\operatorname{stat}\left(\Pi + \sum_{\text{act}} \lambda_N g_N \right), \tag{1.12}$$

where λ_N stands for the Lagrange multipliers, which constitute the set of extra unknowns in the problem. For this price of increasing the number of unknowns one gets the exact fulfillment of the constraint condition (1.9). Each multiplier is once and for all assigned to a given pair of contacting points and its value is updated after each iteration, likewise all other unknowns. These features of the method lead to some purely numerical issues, which must be taken into account. One is paradoxically related to the exact fulfillment of the constraint and requires some modification of the formulation to avoid the division by zero. Another question is a necessity to define the dimension of the matrices in a dynamic way because the real number of unknowns varies during the solution process.

Like in the penalty parameter in the penalty method, the Lagrange multiplier has a physical interpretation, too. It represents a force normal to the contacting surfaces. This value is unknown and does change during the process so it is clearly justified, that it must be considered as an extra unknown in the problem.

Both the presented methods have advantages and disadvantages. It was therefore natural, that some attempts were made to combine the methods to eliminate the disadvantages. This led to the augmented Lagrangian method, where the multipliers do not enter the set of unknowns but their values are updated at each iteration using a certain value of a penalty parameter. Unfortunately, this method requires a complicated numerical algorithm, attributed to Uzawa, including two nested loops. In the outer loop the Lagrange multipliers are kept constant and the update is performed in the inner loop. That is why the method is not commonly used. The majority of computer programs used to analyze contact is based on one of the basic methods, i.e. the penalty method and the Lagrange multiplier method.

The augmented Lagrangian method was successfully applied in the analysis of incompressible bodies (Glowinski and Le Tallec 1984) and scarcely in the analysis of frictionless (Wriggers et al. 1985, Kikuchi and Oden 1988) and frictional contact (Alart and Curnier 1991, Laursen and Simo 1993a as well as Wriggers and Zavarise 1993a).

1.4 The Finite Element Method in Contact Analysis

The constraints in the form (1.6) are in general related to each point lying on the surface of the slave body and its counterpart (orthogonal projection) on the surface of the master body. The finite element discretization of both bodies allows for a discretization of constraints, too. As a consequence, one has to deal with a finite number of additional terms, which must be considered in the modified functional (1.10) or (1.12).

Depending on the character of deformation, to which the contacting bodies are subjected, and the spatial dimension of the problem various types of contact finite elements can be used.

The simplest possible element is the node-to-node element shown in Fig. 1.2a. It can be used in 2D and 3D analysis but its application is restricted only to the cases with small displacements and small strains. This element connects two nodes, one per each contacting body. In this case no distinction between master and slave bodies is necessary.

If problems, where large displacements, especially large relative displacements between the bodies, are treated, node-to-segment and node-to-surface elements should be used to solve 2D and 3D problems, respectively. The elements are presented in Figs. 1.2b and 1.2c. These elements connect one node from the slave body with a number of nodes from the master body.

An element, which is used in the case of contacting beams belongs to a different category (Fig. 1.2d). In this case the distinction between the slave and the master is not introduced, either. The contact points do not coincide with slave nodes but lie between the nodes on both beams. Hence, such an element connects at least two nodes from one body with at least two nodes from the second one.

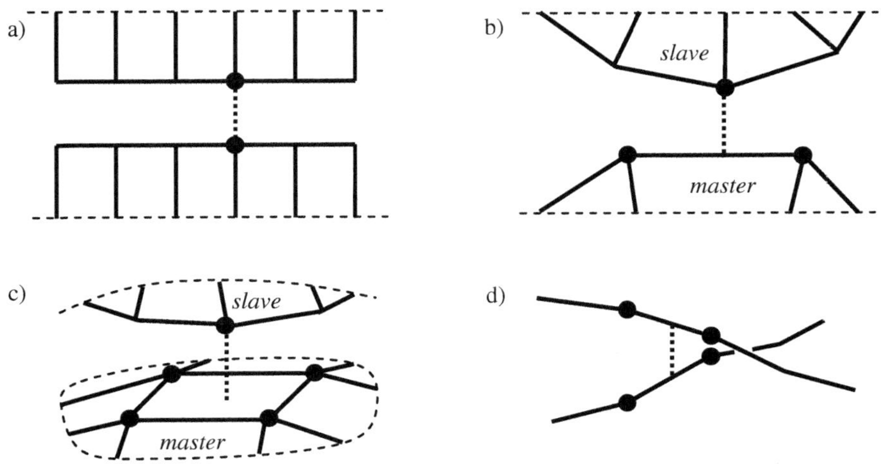

Fig. 1.2 Contact finite elements: a) node-to-node, b) node-to-segment, c) node-to-surface, d) beam-to-beam

Incremental-iterative solution of the functional minimization (1.4) by the finite element method leads at each iteration to a set of equations, which for a pair of separated bodies takes the following matrix form

$$\begin{bmatrix} \mathbf{K}_A & \mathbf{0} \\ \mathbf{0} & \mathbf{K}_B \end{bmatrix} \begin{bmatrix} \Delta\mathbf{q}_A \\ \Delta\mathbf{q}_B \end{bmatrix} + \begin{bmatrix} \mathbf{R}_A \\ \mathbf{R}_B \end{bmatrix} = \begin{bmatrix} \mathbf{0} \\ \mathbf{0} \end{bmatrix}. \tag{1.13}$$

Equation (1.13) actually represents two uncoupled sets of equation, one for each considered bodies.

The situation is different, if additional terms, resulting from the constraint fulfillment using the penalty method (1.10) or the Lagrange multiplier method (1.12), are introduced to the functional (1.4). If one considers the fact, that the penetration function g_N present in the modifications depends on the displacements of the finite element nodes, the related tangent stiffness matrix and the residual vector for the applied contact finite element can be determined. For the case of the penalty method they can be written down in the following form

$$\mathbf{K}_N = \begin{bmatrix} \mathbf{K}_{NAA} & \mathbf{K}_{NAB} \\ \mathbf{K}_{NBA} & \mathbf{K}_{NBB} \end{bmatrix}, \qquad \mathbf{R}_N = \begin{bmatrix} \mathbf{R}_{NA} \\ \mathbf{R}_{NB} \end{bmatrix} \tag{1.14}$$

and for the Lagrange multiplier method as

$$\mathbf{K}_N = \begin{bmatrix} \mathbf{K}_{NAA} & \mathbf{K}_{NAB} & \mathbf{K}_{NA\lambda} \\ \mathbf{K}_{NBA} & \mathbf{K}_{NBB} & \mathbf{K}_{NB\lambda} \\ \mathbf{K}_{N\lambda A} & \mathbf{K}_{N\lambda B} & \mathbf{0} \end{bmatrix}, \qquad \mathbf{R}_N = \begin{bmatrix} \mathbf{R}_{NA} \\ \mathbf{R}_{NB} \\ \mathbf{R}_{N\lambda} \end{bmatrix}. \tag{1.15}$$

If these matrices and vectors are introduced to the finite element formulation, the following coupled set of equation results

$$\begin{bmatrix} (\mathbf{K}_A + \mathbf{K}_{NAA}) & \mathbf{K}_{NAB} \\ \mathbf{K}_{NBA} & (\mathbf{K}_B + \mathbf{K}_{NBB}) \end{bmatrix} \begin{bmatrix} \Delta\mathbf{q}_A \\ \Delta\mathbf{q}_B \end{bmatrix} + \begin{bmatrix} \mathbf{R}_A + \mathbf{R}_{NA} \\ \mathbf{R}_B + \mathbf{R}_{NB} \end{bmatrix} = \begin{bmatrix} \mathbf{0} \\ \mathbf{0} \end{bmatrix} \tag{1.16}$$

for the penalty method and for the Lagrange multiplier method one gets

$$\begin{bmatrix} (\mathbf{K}_A + \mathbf{K}_{NAA}) & \mathbf{K}_{NAB} & \mathbf{K}_{NA\lambda} \\ \mathbf{K}_{NBA} & (\mathbf{K}_B + \mathbf{K}_{NBB}) & \mathbf{K}_{NB\lambda} \\ \mathbf{K}_{N\lambda A} & \mathbf{K}_{N\lambda B} & \mathbf{0} \end{bmatrix} \begin{bmatrix} \Delta\mathbf{q}_A \\ \Delta\mathbf{q}_B \\ \Delta\boldsymbol{\lambda} \end{bmatrix} + \begin{bmatrix} \mathbf{R}_A + \mathbf{R}_{NA} \\ \mathbf{R}_B + \mathbf{R}_{NB} \\ \mathbf{R}_{N\lambda} \end{bmatrix} = \begin{bmatrix} \mathbf{0} \\ \mathbf{0} \\ \mathbf{0} \end{bmatrix}, \tag{1.17}$$

where the unknown vector is extended by the Lagrange multipliers.

Taking contact into account in the finite element analysis leads in a natural way to a connection of the contacting bodies into one entity. For this new body one tangent stiffness matrix and one residual vector can be calculated, since the resulting sets of equations (1.16) or (1.17) are coupled, contrary to (1.13).

1.5 Friction Constraints

Phenomena related to friction have their source at an atomic level. That is why up till now many of the questions concerning friction are not answered. An overview of experimental research on friction can be found in the paper by Woo and Thomas (1980). Numerical methods used in the analysis of friction are discussed in the paper by Oden and Martins (1986). One of the newest contributions to this complicated topic is the monograph by Persson (2000). There are also many papers devoted to micromechanical aspects of friction (e.g. Anand 1993, Stupkiewicz and Mróz 1999 or Stupkiewicz 2001).

In the present work the most commonly used model of dry Coulomb friction is applied. Phenomena requiring a more advanced treatment, like adhesion, lubrication or surface wear are not taken into account.

In the analysis of contact and constraints resulting from a possible relative movement of a contact point on the surface of a contacting body one can distinguish two possible situations. In the first one the friction force is large enough to restrain this movement. In such a case two bodies are stuck one to another and the state is called *stick*. Then, for an arbitrary pair of points $\mathbf{x}_1$ and $\mathbf{x}_{2C}$ shown in Fig. 1.1, the constraint due to the state without sliding can be written in the following form

$$g_T = 0 \, , \tag{1.18}$$

where g_T denotes a relative distance, by which the point $\mathbf{x}_{2C}$ travels on the surface of the master body.

The second case concerns a situation, when the friction force is too small to prevent the relative displacement of the bodies. Such a state is called *slip*. Because the path of movement of an arbitrary point on the surface of the master body is not known apriori, the velocity vector of this point must be integrated in time. A more detailed description of this problem can be found in many contributions, e.g. in the monograph by Wriggers (2002).

Constraint for the slip state can be quite easily formulated, if the friction problem is treated in an analogy to the elastic-plastic material model. Such a concept was suggested in the papers by Bowden and Tabor (1964) as well as by Michałowski and Mróz (1978). In this approach a split of the relative displacement g_T into a part analogous to elastic strain – g_T^e and a part analogous to the plastic strain – g_T^e is introduced

$$g_T = g_T^{\,e} + g_T^{\,p} \, . \tag{1.19}$$

The plastic part corresponding to the real displacement of the contact point can be determined by the above mentioned integration, while the elastic part is subjected to the constraint

$$g_T^{\,e} = 0 \, . \tag{1.20}$$

The constraint (1.18) or (1.20), depending on the nature of the state related to the displacement of the contact point, can be treated by an inclusion of additional terms in the functional in a similar way as the constraints due to the normal contact are treated. In the framework of the penalty method with the penalty parameter ε_T one gets

$$\min\left[\Pi + \sum_{\text{act}}\left(\frac{1}{2}\varepsilon_N g_N^{\,2} + \frac{1}{2}\varepsilon_T g_T^{e\,2}\right)\right] \tag{1.21}$$

and in the Lagrange multiplier method with Lagrange multiplier λ_T –

$$\text{stat}\left[\Pi + \sum_{\text{act}}\left(\lambda_N g_N + \lambda_T g_T^e\right)\right] . \tag{1.22}$$

If the constraint condition is fulfilled exactly, like in the Lagrange multiplier method, this approach corresponds to the rigid-ideally plastic material model. In the case of the penalty method, when the constraint condition is satisfied only with a certain accuracy, this approach constitutes an analogy to a more general elastic-ideally plastic material model, where the penalty parameter ε_T is a counterpart of the elasticity modulus.

The inherent split of the relative displacement into elastic and plastic parts has some justification proved in the experiments by Courtney-Pratt and Eisner (1957) as well as by Anand (1993). They have shown, that the elastic part of the displacement is related to the elastic properties of asperities on the contacting bodies surfaces. In view of this, the penalty parameter ε_T might be considered as a modulus of microelasticity.

The presented method is fully capable of inclusion of hardening or softening phenomena, analogically to the theory of plasticity. If one considers the facts observed in many experiments, that the kinematic friction coefficient is usually smaller than the static one, it should be pointed out that softening should rather be taken into account in this context.

Numerical formulation and solution of the frictional contact with the analogy to plasticity was given in the papers by Wriggers (1987), Wriggers et al. (1990) as well as Laursen and Simo (1993b).

Chapter 2
Frictionless Beam-to-Beam Contact

2.1 Assumptions

Numerical formulation of kinematics and constitutive relations concerning contact can be completely independent of the material model used for the description of contacting bodies. All properties of contact itself are included in additional terms of the functional (1.21) or (1.22), which are determined for active constraints. On the other hand, all the properties related to pure deformations of beams are included in the components of the formula (1.4). Such a statement does not exclude a possibility of a real relation between the properties corresponding to the body interior and to its surface. Nevertheless, in the pure contact analysis these two zones can be treated separately.

However, it is not possible to perform any numerical calculations by the finite element method concerning contact without modeling of the bodies themselves. Hence, the beams in the present considerations must also be analyzed. The literature concerning the beam finite elements is very broad. Some information can be found e.g. in the classical FEM monographs by Zienkiewicz (2000) or by Bathe (1996). The detailed description of this topic lies beyond the scope of the present contribution, so here only basic assumptions are given:

i) the material of which the beams are made is linearly elastic,
ii) the beams undergo large displacements but strains remain small.

A consequence of the second assumption is very important. It comes out, that the beam cross-sections after deformation preserve their shape and size and remain plane. This fact is of a great value, since it simplifies the analysis of contact geometry and the definition of the penetration function.

In the computer programs applied in the numerical analysis of all examples included in this contribution three types of beam finite elements are used. For the beams with rectangular cross-section it is the element based on the updated Lagrangian formulation with polynomial shape functions. This element was described in detail in the paper by Litewka et al. (2001). For the beams with circular cross-section the co-rotational finite element with provision for large rotations, presented by Crisfield (1990) is adopted. Besides, in some examples concerning the electric-mechanical coupling, the classical linear two-node beam finite element with the Hermite's polynomial shape functions is used.

P. Litewka: Finite Element Analysis of Beam-to-Beam Contact, LNACM 53, pp. 13–37.
springerlink.com

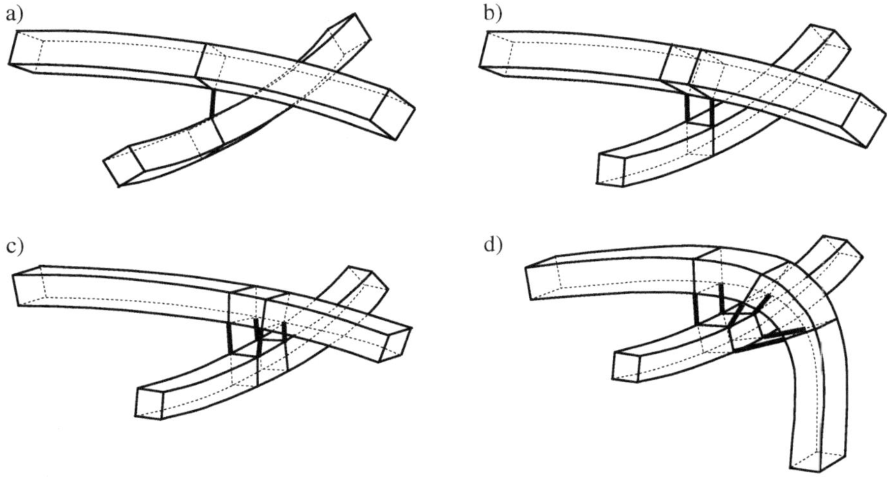

Fig. 2.1 Contact between edges of beams with rectangular cross-sections: a) one pair, b) two pairs, c) four pairs, d) more than four pairs.

The beam-to-beam contact is a specific case, if compared with contact between typical 3D bodies. Due to the proportion between the beam dimensions, where one of them is much larger than the two remaining ones, beams are usually treated as 3D curves represented by their central lines. In the result of this, if one neglects a special case of parallel or conforming beams (curves), the contact between beams can be, at least locally, considered as point-wise. Such an approach is correct for beams with circular cross-sections (Wriggers and Zavarise 1997). In the present considerations another assumption made there is also kept, i.e. the cross-section is small enough to neglect any possible eccentricity of the contact forces acting on the surface of the beam with respect to the beam axis.

In the case of the rectangular cross-section the problem is more complicated. The contact may occur between the curves representing the longitudinal edges of real beams (Litewka and Wriggers 2002a). In this situation several different configurations are possible. They are presented in Fig. 2.1. It is assumed, that the contact is still point-wise but with respect to the edges pairs. In the following it is accepted, that the maximum number of contacting edge pairs is equal to four. This means, that the configuration shown in Fig. 2.1d is not considered. It can be observed, that in such a case at least one of the beams would be wrapped around another, thus undergoing large strains. This would be against the previously introduced assumptions. It can be concluded, that due to even more complicated kinematics and problems with an efficient formulation of a relation between the penetration function and the deformation of the beam axis a full 3D analysis of contact would be much more suitable to treat the case of beams undergoing large strains.

With the assumption of the point-wise contact agreed, several specific issues of discretisation within the finite element method and on the contact finite element formulation itself emerge. Such a case cannot be treated in the same way as contact between solids where pairs: slave node and its orthogonal projection on the master segment or surface are considered. Here the contact points lie between the beam element nodes and only in rare situations, they would coincide. Hence, in the contact search both beams are treated equivalently, no distinction between slave and master is introduced and the contact points must be located simultaneously on both beams.

Hence, the contact finite element resulting from such an approach combines nodes of this pair of beam finite elements, within which the contact points really lie (see Fig. 1.2d).

2.2 Penetration Function

The basis to define the penetration function for two beams in the point-wise contact is finding a pair of two closest points lying on two curves, as shown in Fig. 2.2. In the case of beams with circular cross-sections the points are to belong to curves representing beam axes, while for beams with rectangular cross-sections the points lie on edges. Location of an arbitrary point on the curve is defined by a local curvilinear co-ordinate, ξ_m or ξ_s, respectively. The introduced distinction by means of the subscripts *m* and *s* does not mean the distinction between master and slave beam. It was already mentioned, that in the case of the beam-to-beam contact such a distinction is not necessary – both beams are treated equivalently.

In the global Cartesian system of co-ordinates (x_1, x_2, x_3) each point on the curve is associated with a position vector, $\mathbf{x}_m$ or $\mathbf{x}_s$. Because the penetration function must be determined always in the current configuration, at every iteration of the incrementally treated deformation process these vectors correspond to the current beams configurations. They can be expressed as a sum of the position vector for the point at the initial configuration and the displacement vector

$$\begin{aligned} \mathbf{x}_m &= \mathbf{X}_m + \mathbf{u}_m \,, \\ \mathbf{x}_s &= \mathbf{X}_s + \mathbf{u}_s \,. \end{aligned} \tag{2.1}$$

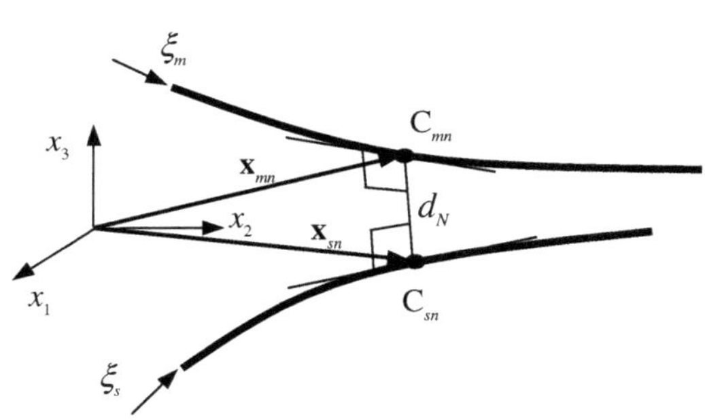

Fig. 2.2 The closest points on two curves

The position vectors $\mathbf{x}_{mn}$ and $\mathbf{x}_{sn}$ of the closest points C_{mn} and C_{sn} on the curves must fulfil the orthogonality conditions between the connecting straight line and the lines tangent to the curves at these points, see Fig. 2.2. These conditions can be written as

$$\begin{aligned}(\mathbf{x}_{mn}-\mathbf{x}_{sn})\circ\mathbf{x}_{mn,m}=0\,,\\ (\mathbf{x}_{mn}-\mathbf{x}_{sn})\circ\mathbf{x}_{sn,s}=0\,,\end{aligned}\quad, \tag{2.2}$$

where the common notation for the partial derivatives with respect to the local co-ordinates ξ_m and ξ_s was introduced:

$$\mathbf{x}_{mn,m}=\frac{\partial\mathbf{x}_{mn}}{\partial\xi_m}\,,$$

$$\mathbf{x}_{sn,s}=\frac{\partial\mathbf{x}_{sn}}{\partial\xi_s}\,.$$

In a general case, when initially straight axes or edges of beams undergo a deformation, the relation (2.2) constitutes a non-linear set of equations. Its solution in the form of a pair of co-ordinates ξ_{mn} and ξ_{sn}, which describe the location of the closest points C_{mn} and C_{sn}, can be found by means of the iterative Newton method. As a starting point for the iterative process a pair of local co-ordinates for any points can be selected. The linearisation of the conditions (2.2) required within the Newton method leads to a set of ordinary linear equations, which allow calculation of the local co-ordinates increments $\Delta\xi_m$ and $\Delta\xi_s$

$$\begin{bmatrix}-\mathbf{x}_{m,m}\circ\mathbf{x}_{m,m}+(\mathbf{x}_m-\mathbf{x}_s)\circ\mathbf{x}_{m,mm} & \mathbf{x}_{m,m}\circ\mathbf{x}_{s,s}\\ -\mathbf{x}_{m,m}\circ\mathbf{x}_{s,s} & \mathbf{x}_{s,s}\circ\mathbf{x}_{s,s}+(\mathbf{x}_m-\mathbf{x}_s)\circ\mathbf{x}_{s,ss}\end{bmatrix}\begin{bmatrix}\Delta\xi_m\\ \Delta\xi_s\end{bmatrix}= \\ =\begin{bmatrix}-(\mathbf{x}_m-\mathbf{x}_s)\circ\mathbf{x}_{m,m}\\ -(\mathbf{x}_m-\mathbf{x}_s)\circ\mathbf{x}_{s,s}\end{bmatrix}. \tag{2.3}$$

All the quantities present in this formula must be determined for the current values of the co-ordinates ξ_m and ξ_s.

Existence and uniqueness of a solution of the equations set (2.2) is a self contained problem. One can easily imagine situations, when the solution does not exist at all or when infinite number of pairs of closest points exist. However, the first possibility can be associated with the case, when the curves represent beams, which cannot get in contact, that is why it does not need to be considered. The second case is related to the situation, when one beam wraps itself around another, similarly as shown in Fig. 2.1d or when beams are parallel or conforming. These configurations are not considered, either, because they are beyond the range set by the assumptions introduced in Section 2.1. A thorough discussion of the closest points projection can be found in the papers by Konyuhov and Schweizerhof (2008, 2009).

With the pair of the closest points C_{mn} and C_{sn} found one can calculate the distance d_N between them (Fig. 2.2)

$$d_N = \left\| \mathbf{x}_{mn} - \mathbf{x}_{sn} \right\| . \tag{2.4}$$

For the beams with rectangular cross-section, where d_N represents the distance between the beam edges, this value is simultaneously equal to the penetration function

$$g_N = d_N \tag{2.5}$$

and in the case of circular cross-sections with radii r_m and r_s the penetration function can be calculated as

$$g_N = d_N - r_m - r_s . \tag{2.6}$$

In the latter case the value of the penetration function can be directly applied in the contact existence criterion

$$\begin{aligned} g_N &= d_N - r_m - r_s \geq 0 \;\Rightarrow\; \text{separate bodies}, \\ g_N &= d_N - r_m - r_s < 0 \;\Rightarrow\; \text{penetration}. \end{aligned} \tag{2.7}$$

In the case of beams with rectangular cross-sections the contact criterion cannot take the same form because the value g_N from the formula (2.5) is always positive. In order to define a special criterion for this case the following vectors, shown in Fig. 2.3, are introduced

i) the vector between the closest points C_{mn} and C_{sn} on the beam edges

$$\mathbf{v}^{\mathrm{CC}} = \mathbf{x}_{sn} - \mathbf{x}_{mn} , \tag{2.8}$$

ii) the vector between the point C_{mn} and the corresponding point A_{mn} located on the beam axis m

$$\mathbf{v}_m^{\mathrm{CA}} = \mathbf{x}_{mA} - \mathbf{x}_{mn} , \tag{2.9}$$

iii) the vector between the point C_{sn} and the corresponding point A_{sn} located on the beam axis s

$$\mathbf{v}_s^{\mathrm{CA}} = \mathbf{x}_{sA} - \mathbf{x}_{sn} . \tag{2.10}$$

It can be noticed in Fig. 2.3, that in the case of separate beams the angles α_m and α_s between these vectors, defined as

$$\begin{aligned} \alpha_m &= \angle\left(\mathbf{v}^{\mathrm{CC}}, \mathbf{v}_m^{\mathrm{CA}}\right), \\ \alpha_s &= \angle\left(-\mathbf{v}^{\mathrm{CC}}, \mathbf{v}_s^{\mathrm{CA}}\right) \end{aligned} \tag{2.11}$$

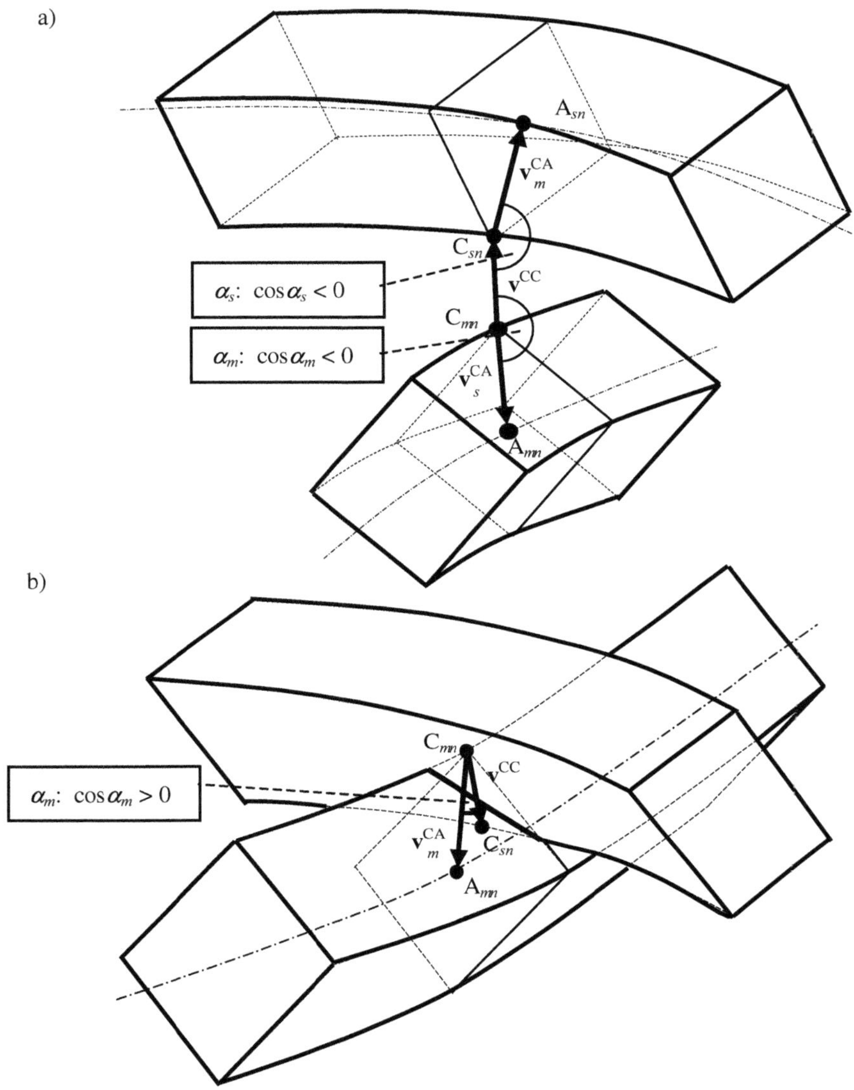

Fig. 2.3 Contact criterion for beams with rectangular cross-sections: a) separation, b) penetration

are obtuse, while in the case of penetration – acute. Hence, the contact criterion for the beams with rectangular cross-sections can be defined as

$$\begin{aligned} \cos\alpha_m < 0 \quad \text{and} \quad \cos\alpha_s < 0 \quad &\Rightarrow \quad \text{separate bodies}, \\ \cos\alpha_m > 0 \quad \text{and} \quad \cos\alpha_s > 0 \quad &\Rightarrow \quad \text{penetration}. \end{aligned} \tag{2.12}$$

2.3 Contact Search

The search of active constraints is usually the most time consuming part of the numerical analysis of contact. It is especially so in the case of contact between 3D solids. If one must also include in the analysis phenomena leading to breaking of bodies into smaller elements or large deformation resulting in contact of a part of a body with its another part (so called *self-contact*), then the problem becomes even more complex. In the finite element method, where the numbers of slave body nodes and master body segments can reach very large values, checking of all possible pairs is absolutely ineffective. To treat these cases methods of a so called hierarchical contact search were developed. In the first step fragments of bodies, which can possibly get in contact are found using very simple algorithms. In the second stage a more thorough local search is performed. Among many papers dealing with this approach one can mention those by Zhong and Nilsson (1989) and by Williams and O'Connor (1995). There are many global search methods and one may choose from; space subdivision (Belytschko and Neal 1991), alternating digital tree (Bonnet and Peraire 1991), heapsorting (Williams and O'Connor 1995), inside-outside algorithm (Wang and Nakamachi 1997), no-binary-search algorithm (Munjiza and Andrews 1998) or augmented spatial digital tree (Feng and Owen 2002). The mostly used local search methods are: node-to-segment (Hallquist 1979) and pinball (Belytschko and Neal 1991). Hierarchical method was also applied in the QuickTrace algorithm (Bruneel and De Rycke 2002) used in the contact search for the case of a cutting tool and a material in a cutting process. The more detailed description of various contact search algorithms can be found in the monograph by Wriggers (2002) or in the paper by Williams and O'Connor (1999).

The problem of the beam-to-beam contact is much easier due to the one-dimensional character of beam geometry and, consequently, a smaller number of involved finite elements. However, even in this simple situation, it is convenient to perform the global search first and after finding a possible contact zone – to carry out a local search to find a location of contact points.

In the first stage of the contact search the closest pair of elements is found (see Fig. 2.4). To this end in every element a point, IC or JC, is introduced, which lies in the centre of the straight line segment connecting the extreme nodes. The position vectors of these fictitious nodes are determined from

$$\begin{aligned} \mathbf{x}_{m\mathrm{IC}} &= \frac{1}{2}\left(\mathbf{x}_{m\mathrm{I}} + \mathbf{x}_{m\mathrm{I+1}}\right), \\ \mathbf{x}_{s\mathrm{JC}} &= \frac{1}{2}\left(\mathbf{x}_{s\mathrm{J}} + \mathbf{x}_{s\mathrm{J+1}}\right). \end{aligned} \tag{2.13}$$

Then the entire set of all elements pairs is searched through to find the one with the smallest distance expressed by

$$d = \left\| \mathbf{x}_{m\mathrm{IC}} - \mathbf{x}_{s\mathrm{JC}} \right\|.$$

Such a search is clearly approximate because the real contact point candidates do not necessarily lie within the elements yielding from this procedure. The stages of local contact search must therefore include a provision to check the adjacent elements, too. However, if a quite reasonable assumption is made, that the length of the beam elements is not smaller than their cross-section dimensions, then the actual contact point candidate lies within the found element or at the most extreme case within one of neighbouring elements.

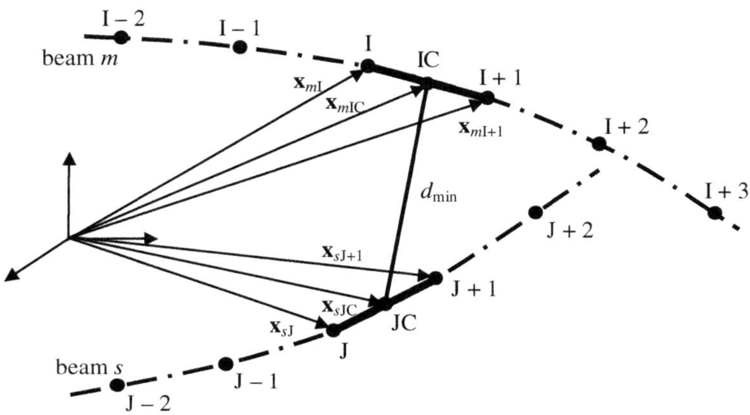

Fig. 2.4 Search for the pair of closest beam elements

The further stages of the contact search must differ for different shapes of the beams cross-sections. For the circular one the situation is relatively simple. In the second stage on the local level the linearisation of the orthogonality conditions (2.3) is applied and using the Newton method the local co-ordinates, ξ_{mn} and ξ_{sn}, of the closest points lying on the beam axes are determined. For the points defined by them to be located within the given beam elements, the following conditions must be fulfilled

$$\begin{aligned} -1 \leq \xi_{mn} \leq 1, \\ -1 \leq \xi_{sn} \leq 1. \end{aligned} \tag{2.14}$$

If any of them is violated than the local search must be repeated for a modified pair of elements, which includes the appropriate adjacent element instead of the one taken in the first attempt. After the pair of points fulfilling the conditions (2.14) is found, the contact criterion (2.7) can be checked. If penetration occurs, then the considered pair is active.

Some more details concerning the approximation of curves representing the axes of beams with circular cross-sections are given in Chapter 4, where methods of 3D curve smoothing using Hermite's polynomials and Bezier's curves are presented.

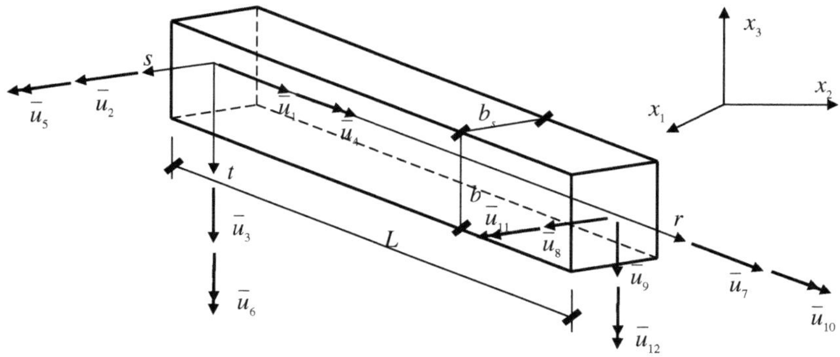

Fig. 2.5 Beam finite element with rectangular cross-section

In the case of beams with rectangular cross-sections the further stages must involve finding the possible contact candidates among beams edges. To this end the co-ordinates of points lying on these edges must be expressed by means of nodal displacements of the beam element, shown in Fig. 2.5. Here, the shape functions for the beam finite element based on updated Lagrangian formulation, presented in the papers by Litewka et al. (2001) and Litewka and Wriggers (2002a) can be used. They are related to the local co-ordinate system (*r*, *s*, *t*) (Fig. 2.5) and have the form:

$$N_1 = 1 - \frac{r}{L}, \; N_2 = \frac{r}{L}, \tag{2.15$_{1-3}$}$$

$$N_3^\alpha = D_\alpha\left(\frac{2r^3}{L^3} - \frac{3r^2}{L^2} - \frac{d_\alpha r}{L} + d_\alpha + 1\right), \; N_4^\alpha = D_\alpha\left(-\frac{2r^3}{L^3} + \frac{3r^2}{L^2} + \frac{d_\alpha r}{L}\right),$$

$$N_5^\alpha = \frac{D_\alpha L}{2}\left[\frac{2r^3}{L^3} - \frac{(4+d_\alpha)r^2}{L^2} + \frac{(2+d_\alpha)r}{L}\right], \; N_7^\alpha = \frac{6D_\alpha}{L}\left(-\frac{r^2}{L^2} + \frac{r}{L}\right),$$

$$N_6^\alpha = \frac{D_\alpha L}{2}\left[\frac{2r^3}{L^3} - \frac{(2-d_\alpha)r^2}{L^2} - \frac{d_\alpha r}{L}\right], \; N_9^\alpha = D_\alpha\left[\frac{3r^2}{L^2} - \frac{(4-d_\alpha)r}{L}\right],$$

$$N_8^\alpha = D_\alpha\left[\frac{3r^2}{L^2} - \frac{(4+d_\alpha)r}{L} + d_\alpha + 1\right], \tag{2.15$_{4,5}$}$$

where

$$d_\alpha = \frac{12\kappa E I_\beta}{GAL^2}, \; D_\alpha = \frac{1}{1+d_\alpha}, \; \alpha = s \text{ or } t, \; \beta = t \text{ or } s,$$

L is the element length, E and G denote the Young's and Kirchhoff's moduli, respectively, I_s and I_t – the second moments of area of the cross-section with respect to the axes s and t, A – the cross-section area and κ – the shear correction factor.

Within the assumptions discussed in Section 2.1 it can be stated, that the beam cross-section shown in Fig. 2.6 undergoes the rigid body movement with three components of displacements and three rotations. Then three components of the displacement of a point C, which is a corner of the cross-section can be expressed by:

$$
\begin{aligned}
\bar{u}_{r\mathrm{C}} &= \bar{u}_{r\mathrm{A}} + \bar{u}_r(\bar{\varphi}) = \left(N_1\bar{u}_1 + N_2\bar{u}_7\right) \\
&\quad + \left[\frac{t\bar{\varphi}_s - s\bar{\varphi}_t}{\bar{\varphi}}\sin\bar{\varphi} + \frac{\bar{\varphi}_r\left(s\bar{\varphi}_s + t\bar{\varphi}_t\right)}{\bar{\varphi}^2}\left(1-\cos\bar{\varphi}\right)\right], \\
\bar{u}_{s\mathrm{C}} &= \bar{u}_{s\mathrm{A}} + \bar{u}_s(\bar{\varphi}) = \left(N_3^s\bar{u}_2 + N_4^s\bar{u}_8 + N_5^s\bar{u}_6 + N_6^s\bar{u}_{12}\right) + \\
&\quad + \left[-\frac{t\bar{\varphi}_r}{\bar{\varphi}}\sin\bar{\varphi} - \frac{s\left(\bar{\varphi}_r^{\,2} + \bar{\varphi}_t^{\,2}\right) - t\bar{\varphi}_s\bar{\varphi}_t}{\bar{\varphi}^2}\left(1-\cos\bar{\varphi}\right)\right], \\
\bar{u}_{t\mathrm{C}} &= \bar{u}_{t\mathrm{A}} + \bar{u}_t(\bar{\varphi}) = \left(N_3^t\bar{u}_3 + N_4^t\bar{u}_9 - N_5^t\bar{u}_5 - N_6^t\bar{u}_{11}\right) + \\
&\quad + \left[\frac{s\bar{\varphi}_r}{\bar{\varphi}}\sin\bar{\varphi} - \frac{t\left(\bar{\varphi}_r^{\,2} + \bar{\varphi}_s^{\,2}\right) - s\bar{\varphi}_s\bar{\varphi}_t}{\bar{\varphi}^2}\left(1-\cos\bar{\varphi}\right)\right],
\end{aligned}
\tag{2.16}
$$

where the following simplifying notation was adopted:

$$
\begin{aligned}
\bar{\varphi}_r &= N_1\bar{u}_4 + N_2\bar{u}_{10}\,, \\
\bar{\varphi}_s &= N_7^t\bar{u}_3 - N_7^t\bar{u}_9 + N_8^t\bar{u}_5 + N_9^t\bar{u}_{11}\,, \\
\bar{\varphi}_t &= -N_7^s\bar{u}_2 + N_7^s\bar{u}_8 + N_8^s\bar{u}_6 + N_9^s\bar{u}_{12}\,, \\
\bar{\varphi} &= \sqrt{\left(\bar{\varphi}_r^{\,2} + \bar{\varphi}_s^{\,2} + \bar{\varphi}_t^{\,2}\right)}.
\end{aligned}
$$

To determine displacements of a point C on the beam edge using formulae (2.16) one must substitute for the co-ordinates s and t the appropriate values:

$$
s = \pm\frac{b_s}{2}, \quad t = \pm\frac{b_t}{2}.
$$

Substitution of $s = t = 0$ allows for calculation of the displacements of a point A (Fig. 2.6) located on the beam axis.

Displacements yielding from the formulae (2.16) are expressed in the local co-ordinate system (r, s, t). To get the displacements expressed in the global system (x_1, x_2, x_3), the transformation must be carried out. To this end the initial transformation matrix $\mathbf{T}$ (Litewka et al. 2001) can be used

$$
\mathbf{u} = \left(u_1, u_2, u_2\right)^T = \mathbf{T}^T\left(\bar{u}_r, \bar{u}_s, \bar{u}_t\right)^T \tag{2.17}
$$

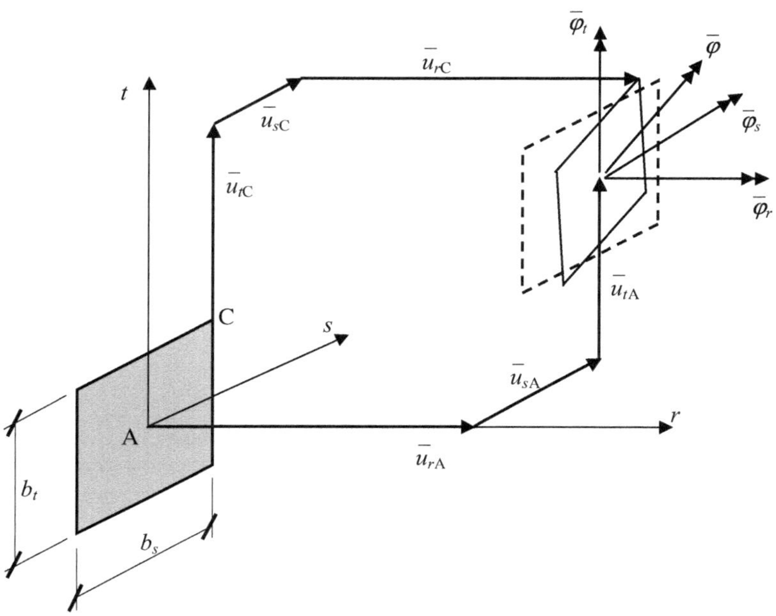

Fig. 2.6 Displacements of a rectangular cross-section of a beam in local co-ordinate system

If it is assumed that the beam finite element is initially straight, then the position vector of any point on the beam edge can be given in the form

$$\mathbf{x} = \mathbf{X} + \mathbf{u} = \left(N_1\mathbf{X}^{\mathbf{I}} + N_2\mathbf{X}^{\mathbf{I+1}}\right) + \left(u_1, u_2, u_3\right)^T \tag{2.18}$$

where $\mathbf{X}^{\mathrm{I}}$ and $\mathbf{X}^{\mathrm{I+1}}$ are the initial position vectors of the beam element nodes.

Having this formulation of the position vectors of the points on the beam edges related to the nodal displacements one can find for the beam pair the edges, which are contact candidates. First, the linearisation of the orthogonality conditions (2.3) and the Newton method are used to perform the search for the closest points between four edges of beam m and the axis of beam s. From these four pairs and the respective values of the distance between the closest points, two closest edges from beam m – $m1$ and $m2$, shown in Fig. 2.7, are selected. It must also be noted, that the linearisation of the orthogonality conditions (2.3) is expressed for the dimensionless local co-ordinates ξ_m and ξ_s, while the displacements in Eq. (2.18) are expressed by the local co-ordinates r_m and r_s. So the following relations also must be taken into account:

$$\xi_m = \frac{2r_m}{L_m} - 1\ ,\ \ \xi_s = \frac{2r_s}{L_s} - 1\,. \tag{2.19}$$

In the next step the roles of the beams are exchanged and two edges, $s1$ and $s2$, which are the closest edges of the beam s with respect to the axis of the beam m, are found.

In the last stage of the contact search the closest points are found for four pairs of the edges $m1 - s1$, $m1 - s2$, $m2 - s1$ and $m2 - s2$. For all these pairs the contact criterion (2.12) is checked.

During each run of the procedure of the closest points location it must be checked, if the located points lie within the considered elements. Again, if any of the conditions (2.14) is violated, then the corresponding closest point search must be repeated using the adequate neighbouring element.

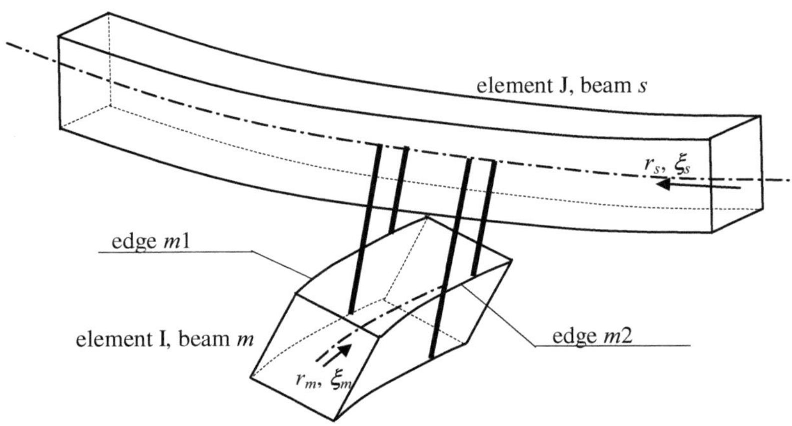

Fig. 2.7 Search of the closest edges

2.4 Weak Form and Kinematic Variables for Contact

Minimization of the functional (1.10) or finding the stationary point for the functional (1.12) lead to the same condition, that the first variation of the functional must equal to zero. This corresponds to the principle of virtual work. In the penalty method one gets

$$\delta\left(\Pi + \sum_{\text{akt}} \frac{1}{2} \varepsilon_N g_N{}^2\right) = 0 \tag{2.20}$$

and in the Lagrange multipliers method –

$$\delta\left(\Pi + \sum_{\text{akt}} \lambda_N g_N\right) = 0\,. \tag{2.21}$$

For a single active pair the additional component of the functional variation can be expressed as

$$
\begin{aligned}
\delta\Pi_N^{\varepsilon} &= \delta\left(\frac{1}{2}\varepsilon_N g_N{}^2\right) = \varepsilon_N g_N \delta g_N \,, \\
\delta\Pi_N^{\lambda} &= \delta(\lambda_N g_N) = \lambda_N \delta g_N + \delta\lambda_N g_N \,,
\end{aligned} \tag{2.22}
$$

where the superscript ε relates to the penalty method and λ – to the Lagrange multipliers method. Expressions (2.22) constitute the elements of the weak formulation of frictionless contact between beams. To solve this strongly non-linear problem the incremental-iterative Newton-Raphson method can be applied. Its use requires calculation of the linearisation of the weak form. The linearisation of the additional contact terms leads to the following formulae:

$$
\begin{aligned}
\Delta\delta\Pi_N^{\varepsilon} &= \varepsilon_N \Delta g_N \delta g_N + \varepsilon_N g_N \Delta\delta g_N \,, \\
\Delta\delta\Pi_N^{\lambda} &= \delta\lambda_N \Delta g_N + \Delta\lambda_N \delta g_N + \lambda_N \Delta\delta g_N \,.
\end{aligned} \tag{2.23}
$$

After the discretisation within the finite element method formulae (2.22) and (2.23) provide the residual vector and the tangent stiffness matrix, respectively, for the contact finite element. To find the explicit form of this two quantities it is necessary to calculate the variation, the linearisation and the linearisation of the variation of the penetration function g_N expressed by (2.5) or (2.6). To this end one uses the variation and the linearisation of the position vectors for the current contact points

$$
\begin{aligned}
\delta\mathbf{x}_{mn} &= \mathbf{x}_{mn,m}\delta\xi_{mn} + \delta\mathbf{u}_{mn} \,, \\
\delta\mathbf{x}_{sn} &= \mathbf{x}_{sn,s}\delta\xi_{sn} + \delta\mathbf{u}_{sn} \,, \\
\Delta\mathbf{x}_{mn} &= \mathbf{x}_{mn,m}\Delta\xi_{mn} + \Delta\mathbf{u}_{mn} \,, \\
\Delta\mathbf{x}_{sn} &= \mathbf{x}_{sn,s}\Delta\xi_{sn} + \Delta\mathbf{u}_{sn} \,.
\end{aligned} \tag{2.24}
$$

After some basic operations the necessary kinematic variables for the normal contact can be expressed as (Wriggers and Zavarise, 1997)

$$
\begin{aligned}
\Delta g_N &= (\Delta\mathbf{u}_{mn} - \Delta\mathbf{u}_{sn}) \circ \mathbf{n} \,, \\
\delta g_N &= (\delta\mathbf{u}_{mn} - \delta\mathbf{u}_{sn}) \circ \mathbf{n} \,,
\end{aligned}
$$

$$
\begin{aligned}
\Delta\delta g_N = {} & (\Delta\delta\mathbf{u}_{mn} - \Delta\delta\mathbf{u}_{sn}) \circ \mathbf{n} + (\delta\mathbf{u}_{mn,m}\Delta\xi_{mn} - \delta\mathbf{u}_{sn,s}\Delta\xi_{sn}) \circ \mathbf{n} + \\
& + (\Delta\mathbf{u}_{mn,m}\delta\xi_{mn} - \Delta\mathbf{u}_{sn,s}\delta\xi_{sn}) \circ \mathbf{n} + \\
& + (\mathbf{x}_{mn,mm}\Delta\xi_{mn}\delta\xi_{mn} - \mathbf{x}_{sn,ss}\Delta\xi_{sn}\delta\xi_{sn}) \circ \mathbf{n} + \\
& + \frac{1}{g_N}(\delta\mathbf{u}_{mn} + \mathbf{x}_{mn,m}\delta\xi_{mn} - \delta\mathbf{u}_{sn} - \mathbf{x}_{sn,s}\delta\xi_{sn}) \\
& \times (\mathbf{1} - \mathbf{n} \otimes \mathbf{n})(\Delta\mathbf{u}_{mn} + \mathbf{x}_{mn,m}\Delta\xi_{mn} - \Delta\mathbf{u}_{sn} - \mathbf{x}_{sn,s}\Delta\xi_{sn}).
\end{aligned} \tag{2.25}
$$

During the derivation of (2.25) it must be taken into account, that the position vectors of the contact points on the curves are functions of the local co-ordinates ξ_m and ξ_s. These formulae contain a unit normal vector defined as

$$\mathbf{n} = \frac{\left(\mathbf{x}_{mn} - \mathbf{x}_{sn}\right)}{d_N} = \frac{\mathbf{x}_{mn,m} \times \mathbf{x}_{sn,s}}{\left\|\mathbf{x}_{mn,m} \times \mathbf{x}_{sn,s}\right\|} . \tag{2.26}$$

It must also be noted, that all the expressions in (2.25) and (2.26) must be determined at the current positions of the contact points C_{mn} and C_{sn}, what is emphasized with an additional subscript n.

The formula $(2.25)_3$ involves two rather unusual variables: $\Delta\delta\mathbf{u}_{mn}$ and $\Delta\delta\mathbf{u}_{sn}$, which in the typical finite element formulations are zero. It is so because usually a displacement approximation is linearly dependent on nodal displacements, so its second partial derivatives vanish. In the contact analysis for beams with rectangular cross-sections the relation (2.18) is not linear and the above mentioned variables are not zero. Similar situation occurs in some methods of curve smoothing, what is discussed in Chapter 4.

There is one more problem with the formula $(2.25)_3$. It involves a component, which is divided by the distance g_N. When the Lagrange multipliers method is used, what leads to the exact fulfillment of the constraint condition (1.9), this component may cause numerical problems due to division by a very small number. To avoid this difficulty one can introduce a different definition of the normal vector, which includes a vector product of tangent vectors, as shown by the second equality in the formula (2.26). Thanks to this change the discussed component of the formula for $\Delta\delta g_N$ includes division by the length of this vector product and the singularity is eliminated.

For the explicit expression of kinematic variables the variation and the linearisation of the local co-ordinates $\delta\xi_{mn}$, $\delta\xi_{sn}$, $\Delta\xi_{mn}$ and $\Delta\xi_{sn}$ must be derived, too. They can be obtained from the linearisation and the variation of the orthogonality conditions (2.2). Each of this operations leads to a system of two equations in the following form

$$\begin{bmatrix} \Delta\xi_{mn} \\ \Delta\xi_{sn} \end{bmatrix} = \begin{bmatrix} \mathbf{x}_{mn,m} \circ \mathbf{x}_{mn,m} + \mathbf{x}_{ms} \circ \mathbf{x}_{mn,mm} & -\mathbf{x}_{mn,m} \circ \mathbf{x}_{sn,s} \\ \mathbf{x}_{mn,m} \circ \mathbf{x}_{sn,s} & -\mathbf{x}_{sn,s} \circ \mathbf{x}_{sn,s} + \mathbf{x}_{ms} \circ \mathbf{x}_{sn,ss} \end{bmatrix}^{-1}$$
$$\times \left(\begin{bmatrix} -\mathbf{x}_{mn,m}^{T} & \mathbf{x}_{mn,m}^{T} \\ -\mathbf{x}_{sn,s}^{T} & \mathbf{x}_{sn,s}^{T} \end{bmatrix} \begin{bmatrix} \Delta\mathbf{u}_{mn} \\ \Delta\mathbf{u}_{sn} \end{bmatrix} - \begin{bmatrix} \mathbf{x}_{ms}^{T} & \mathbf{0} \\ \mathbf{0} & \mathbf{x}_{ms}^{T} \end{bmatrix} \begin{bmatrix} \Delta\mathbf{u}_{mn,m} \\ \Delta\mathbf{u}_{sn,s} \end{bmatrix} \right),$$

which can be solved to express the pairs of unknowns. These equations can be put more briefly as

$$\begin{bmatrix} \Delta\xi_{mn} \\ \Delta\xi_{sn} \end{bmatrix} = \mathbf{A}^{-1} \left(\mathbf{B} \begin{bmatrix} \Delta\mathbf{u}_{mn} \\ \Delta\mathbf{u}_{sn} \end{bmatrix} + \mathbf{C} \begin{bmatrix} \Delta\mathbf{u}_{mn,m} \\ \Delta\mathbf{u}_{sn,s} \end{bmatrix} \right) , \tag{2.27}$$

where the matrices **A**, **B** and **C** as well as the vector $\mathbf{x}_{ms} = \mathbf{x}_{mn} - \mathbf{x}_{sn}$ were introduced to simplify the notation.

Similarly, the variations of the local co-ordinates follow from

$$\begin{bmatrix} \delta\xi_{mn} \\ \delta\xi_{sn} \end{bmatrix} = \mathbf{A}^{-1}\left(\mathbf{B}\begin{bmatrix} \delta\mathbf{u}_{mn} \\ \delta\mathbf{u}_{sn} \end{bmatrix} + \mathbf{C}\begin{bmatrix} \delta\mathbf{u}_{mn,m} \\ \delta\mathbf{u}_{sn,s} \end{bmatrix}\right) . \tag{2.28}$$

2.5 Discretisation of Kinematic Variables

All the kinematic variables derived in Section 2.4 can now be expressed in terms of nodal displacements of two beam elements, within which the current contact points are located. These nodal displacements in respective element local co-ordinates (see Fig. 2.5) can be grouped into the following vectors

$$\begin{aligned} \bar{\mathbf{u}}_M &= \left(\bar{u}_{1m},\ \bar{u}_{2m},\ ...,\ \bar{u}_{12m}\right)^T , \\ \bar{\mathbf{u}}_S &= \left(\bar{u}_{1s},\ \bar{u}_{2s},\ ...,\ \bar{u}_{12s}\right)^T . \end{aligned} \tag{2.29}$$

Firstly, the variation and the linearisation of the displacement vectors $\mathbf{u}_{mn}$ and $\mathbf{u}_{sn}$ as well as their derivatives with respect to the local co-ordinates ξ_m and ξ_s are presented. For the values in the local co-ordinate system (r, s, t) one can write down

$$\begin{aligned} \Delta\bar{\mathbf{u}}_{mn} &= \frac{\partial\bar{\mathbf{u}}_{mn}}{\partial\bar{\mathbf{u}}_M}\Delta\bar{\mathbf{u}}_M = \mathbf{D}_{mn}\Delta\bar{\mathbf{u}}_M , \\ \Delta\bar{\mathbf{u}}_{sn} &= \frac{\partial\bar{\mathbf{u}}_{sn}}{\partial\bar{\mathbf{u}}_S}\Delta\bar{\mathbf{u}}_S = \mathbf{D}_{sn}\Delta\bar{\mathbf{u}}_S , \end{aligned} \tag{2.30}$$

$$\begin{aligned} \delta\bar{\mathbf{u}}_{mn} &= \frac{\partial\bar{\mathbf{u}}_{mn}}{\partial\bar{\mathbf{u}}_M}\delta\bar{\mathbf{u}}_M = \mathbf{D}_{mn}\delta\bar{\mathbf{u}}_M , \\ \delta\bar{\mathbf{u}}_{sn} &= \frac{\partial\bar{\mathbf{u}}_{sn}}{\partial\bar{\mathbf{u}}_S}\delta\bar{\mathbf{u}}_S = \mathbf{D}_{sn}\delta\bar{\mathbf{u}}_S , \end{aligned} \tag{2.31}$$

$$\begin{aligned} \Delta\bar{\mathbf{u}}_{mn,m} &= \frac{\partial\bar{\mathbf{u}}_{mn,m}}{\partial\bar{\mathbf{u}}_M}\Delta\bar{\mathbf{u}}_M = \mathbf{E}_{mn}\Delta\bar{\mathbf{u}}_M , \\ \Delta\bar{\mathbf{u}}_{sn,s} &= \frac{\partial\bar{\mathbf{u}}_{sn,s}}{\partial\bar{\mathbf{u}}_S}\Delta\bar{\mathbf{u}}_S = \mathbf{E}_{sn}\Delta\bar{\mathbf{u}}_S , \end{aligned} \tag{2.32}$$

$$\begin{aligned} \delta\bar{\mathbf{u}}_{mn,m} &= \frac{\partial\bar{\mathbf{u}}_{mn,m}}{\partial\bar{\mathbf{u}}_M}\delta\bar{\mathbf{u}}_M = \mathbf{E}_{mn}\delta\bar{\mathbf{u}}_M , \\ \delta\bar{\mathbf{u}}_{sn,s} &= \frac{\partial\bar{\mathbf{u}}_{sn,s}}{\partial\bar{\mathbf{u}}_S}\delta\bar{\mathbf{u}}_S = \mathbf{E}_{sn}\delta\bar{\mathbf{u}}_S . \end{aligned} \tag{2.33}$$

The matrices $\mathbf{D}_{mn}$, $\mathbf{D}_{sn}$, $\mathbf{E}_{mn}$ and $\mathbf{E}_{sn}$, all of dimensions (3×12), can be derived by tedious differentiation of the displacement approximation (2.18) with respect to the local co-ordinates. Results of these calculations are presented in Appendix 1.

Transformation of the expressions (2.30) – (2.33) from the local co-ordinate systems to the global co-ordinates (x_1, x_2, x_3) yields the following relations

$$\begin{aligned}\Delta\mathbf{u}_{mn} &= \left(\mathbf{T}_m{}^T\mathbf{D}_{mn}\mathbf{T}_{12m}\right)\Delta\mathbf{u}_M = \mathbf{G}_{mn}\Delta\mathbf{u}_M\ ,\\ \Delta\mathbf{u}_{sn} &= \left(\mathbf{T}_s{}^T\mathbf{D}_{sn}\mathbf{T}_{12s}\right)\Delta\mathbf{u}_S = \mathbf{G}_{sn}\Delta\mathbf{u}_S\ ,\end{aligned} \tag{2.34}$$

$$\begin{aligned}\delta\mathbf{u}_{mn} &= \mathbf{G}_{mn}\delta\mathbf{u}_M\ ,\\ \delta\mathbf{u}_{sn} &= \mathbf{G}_{sn}\delta\mathbf{u}_S\ ,\end{aligned} \tag{2.35}$$

$$\begin{aligned}\Delta\mathbf{u}_{mn,m} &= \left(\mathbf{T}_m{}^T\mathbf{E}_{mn}\mathbf{T}_{12m}\right)\Delta\mathbf{u}_M = \mathbf{H}_{mn}\Delta\mathbf{u}_M\ ,\\ \Delta\mathbf{u}_{sn,s} &= \left(\mathbf{T}_s{}^T\mathbf{E}_{sn}\mathbf{T}_{12s}\right)\Delta\mathbf{u}_S = \mathbf{H}_{sn}\Delta\mathbf{u}_S\ ,\end{aligned} \tag{2.36}$$

$$\begin{aligned}\delta\mathbf{u}_{mn,m} &= \mathbf{H}_{mn}\delta\mathbf{u}_M\ ,\\ \delta\mathbf{u}_{sn,s} &= \mathbf{H}_{sn}\delta\mathbf{u}_S\ ,\end{aligned} \tag{2.37}$$

where the additional transformation matrix is defined by (Litewka et al. 2001)

$$\mathbf{T}_{12\alpha} = \begin{bmatrix}\mathbf{T}_\alpha & \mathbf{0} & \mathbf{0} & \mathbf{0}\\ \mathbf{0} & \mathbf{T}_\alpha & \mathbf{0} & \mathbf{0}\\ \mathbf{0} & \mathbf{0} & \mathbf{T}_\alpha & \mathbf{0}\\ \mathbf{0} & \mathbf{0} & \mathbf{0} & \mathbf{T}_\alpha\end{bmatrix},\quad \alpha = m \text{ or } s. \tag{2.38}$$

The linearisation of the local co-ordinates of the contact points can be expressed by means of the nodal displacements as

$$\begin{bmatrix}\Delta\xi_{mn}\\ \Delta\xi_{sn}\end{bmatrix} = \mathbf{A}^{-1}\left\{\mathbf{B}\begin{bmatrix}\mathbf{G}_{mn} & \mathbf{0}\\ \mathbf{0} & \mathbf{G}_{sn}\end{bmatrix} + \mathbf{C}\begin{bmatrix}\mathbf{H}_{mn} & \mathbf{0}\\ \mathbf{0} & \mathbf{H}_{sn}\end{bmatrix}\right\}\begin{bmatrix}\Delta\mathbf{u}_M\\ \Delta\mathbf{u}_S\end{bmatrix}$$

and substitution of the matrices **B** and **C** following from Eq. (2.27) yields the matrix form

$$\begin{aligned}\begin{bmatrix}\Delta\xi_{mn}\\ \Delta\xi_{sn}\end{bmatrix} &= \mathbf{A}^{-1}\begin{bmatrix}\left(-\mathbf{x}_{mn,m}{}^T\mathbf{G}_{mn} - \mathbf{x}_{ms}{}^T\mathbf{H}_{mn}\right) & \mathbf{x}_{mn,m}{}^T\mathbf{G}_{sn}\\ -\mathbf{x}_{sn,s}{}^T\mathbf{G}_{mn} & \left(\mathbf{x}_{sn,s}{}^T\mathbf{G}_{sn} - \mathbf{x}_{ms}{}^T\mathbf{H}_{sn}\right)\end{bmatrix}\begin{bmatrix}\Delta\mathbf{u}_M\\ \Delta\mathbf{u}_S\end{bmatrix} =\\ &= \mathbf{F}\begin{bmatrix}\Delta\mathbf{u}_M\\ \Delta\mathbf{u}_S\end{bmatrix}.\end{aligned} \tag{2.39}$$

Similarly, the variations of the local co-ordinates are given by

$$\begin{bmatrix} \delta\xi_{mn} \\ \delta\xi_{sn} \end{bmatrix} = \mathbf{F} \begin{bmatrix} \delta\mathbf{u}_M \\ \delta\mathbf{u}_S \end{bmatrix}. \tag{2.40}$$

Furthermore, to simplify the notation the following matrix is introduced

$$\mathbf{L} = [\mathbf{G}_{mn} \quad -\mathbf{G}_{sn}] + [\mathbf{x}_{mn,m} \quad -\mathbf{x}_{sn,s}]\mathbf{F}. \tag{2.41}$$

2.6 Residual Vector and Tangent Stiffness Matrix

Having discretised the kinematic variables, as presented in Section 2.5, finally the residual vector and the tangent stiffness matrix for the contact element can be calculated for both formulations: the penalty method and the Lagrange multipliers method. The additional components (2.22) in the functional variation can be expressed in the following form

$$\begin{aligned} \delta\Pi_N^{\varepsilon} &= \left(\delta\mathbf{u}_M{}^T, \delta\mathbf{u}_S{}^T\right)\left[\mathbf{R}_N^{\varepsilon}\right]_{24\times 1}, \\ \delta\Pi_N^{\lambda} &= \left(\delta\mathbf{u}_M{}^T, \delta\mathbf{u}_S{}^T, \delta\lambda_N\right)\left[\mathbf{R}_N^{\lambda}\right]_{25\times 1}, \end{aligned} \tag{2.42}$$

where the respective notation for the residual vectors in both methods was introduced. And the linearisation (2.23) yields the following

$$\begin{aligned} \Delta\delta\Pi_N^{\varepsilon} &= \left(\delta\mathbf{u}_M{}^T, \delta\mathbf{u}_S{}^T\right)\left[\mathbf{K}_N^{\varepsilon}\right]_{24\times 24}\left(\Delta\mathbf{u}_M, \Delta\mathbf{u}_S\right)^T, \\ \Delta\delta\Pi_N^{\lambda} &= \left(\delta\mathbf{u}_M{}^T, \delta\mathbf{u}_S{}^T, \delta\lambda_N\right)\left[\mathbf{K}_N^{\lambda}\right]_{25\times 25}\left(\Delta\mathbf{u}_M, \Delta\mathbf{u}_S, \Delta\lambda_N\right)^T \end{aligned} \tag{2.43}$$

with the appropriate notation for the tangent stiffness matrices for both methods.

The residual vector for the penalty method takes the form

$$\mathbf{R}_N^{\varepsilon} = \varepsilon_N g_N \mathbf{R}_1, \tag{2.44}$$

where

$$\mathbf{R}_1 = \begin{bmatrix} \mathbf{G}_{mn}{}^T\mathbf{n} \\ -\mathbf{G}_{sn}{}^T\mathbf{n} \end{bmatrix}, \tag{2.45}$$

while for the Lagrange multipliers method one gets

$$\mathbf{R}_N^{\lambda} = \begin{bmatrix} \lambda_N \mathbf{R}_1 \\ g_N \end{bmatrix}. \tag{2.46}$$

The tangent stiffness matrix in the penalty method can be written down as

$$\mathbf{K}_N^{\varepsilon} = \varepsilon_N \mathbf{K}_1 + \varepsilon_N g_N \mathbf{K}_2 \tag{2.47}$$

and its submatrices are

$$\mathbf{K}_1 = \mathbf{R}_1 \otimes \mathbf{R}_1 ,$$

$$\mathbf{K}_2 = \begin{bmatrix} \mathbf{H}_{mn}{}^T\mathbf{n} & \mathbf{0} \\ \mathbf{0} & -\mathbf{H}_{sn}{}^T\mathbf{n} \end{bmatrix}\mathbf{F} + \mathbf{F}^T\begin{bmatrix} \mathbf{n}^T\mathbf{H}_{mn} & \mathbf{0} \\ \mathbf{0} & -\mathbf{n}^T\mathbf{H}_{sn} \end{bmatrix} + \begin{bmatrix} \mathbf{G}_{nm} & \mathbf{0} \\ \mathbf{0} & -\mathbf{G}_{ns} \end{bmatrix} + \\ + \mathbf{F}^T\begin{bmatrix} \mathbf{n}^T\mathbf{x}_{mn,mm} & \mathbf{0} \\ \mathbf{0} & -\mathbf{n}^T\mathbf{x}_{sn,ss} \end{bmatrix}\mathbf{F} + \frac{1}{g_N}\mathbf{L}^T\left(\mathbf{1} - \mathbf{n}\cdot\mathbf{n}^T\right)\mathbf{L}\Big\} . \tag{2.48}$$

For the Lagrange multipliers method the following matrix is obtained

$$\mathbf{K}_N^\lambda = \begin{bmatrix} \lambda_N \mathbf{K}_2 & \mathbf{R}_1 \\ \mathbf{R}_1{}^T & \mathbf{0} \end{bmatrix} . \tag{2.49}$$

In the matrix $\mathbf{K}_2$ expressed in $(2.48)_2$ two yet undefined matrices $\mathbf{G}_{nm}$ and $\mathbf{G}_{ns}$ are present. They result from the expressions $\Delta\delta\mathbf{u}_{mn}$ and $\Delta\delta\mathbf{u}_{sn}$, respectively, included in Eq. $(2.25)_3$ defining $\Delta\delta g_N$. For the simplicity of the notation each of the matrices $\mathbf{G}_{mn}$ and $\mathbf{G}_{sn}$ is split into three 12-component vectors

$$\mathbf{G}_{mn} = \left[\mathbf{G}_{1mn}, \mathbf{G}_{2mn}, \mathbf{G}_{3mn}\right]^T , \\ \mathbf{G}_{sn} = \left[\mathbf{G}_{1sn}, \mathbf{G}_{2sn}, \mathbf{G}_{3sn}\right]^T \tag{2.50}$$

and the unit normal vector is given by its components

$$\mathbf{n} = \left[n_1, n_2, n_3\right]^T . \tag{2.51}$$

Now the entries in the matrices $\mathbf{G}_{nm}$ and $\mathbf{G}_{ns}$ can be expressed as

$$\mathbf{G}_{nm} = n_1\mathbf{G}_{d1mn} + n_2\mathbf{G}_{d2mn} + n_3\mathbf{G}_{d3mn} , \\ \mathbf{G}_{ns} = n_1\mathbf{G}_{d1sn} + n_2\mathbf{G}_{d2sn} + n_3\mathbf{G}_{d3sn} , \tag{2.52}$$

where the submatrices result from the reformulation of the variables $\Delta\delta\mathbf{u}_{mn}$ and $\Delta\delta\mathbf{u}_{sn}$ as

$$\Delta\delta\mathbf{u}_{mn} = \begin{bmatrix} \delta\mathbf{u}_M{}^T\,\mathbf{G}_{d1mn}\,\Delta\mathbf{u}_M \\ \delta\mathbf{u}_M{}^T\,\mathbf{G}_{d2mn}\,\Delta\mathbf{u}_M \\ \delta\mathbf{u}_M{}^T\,\mathbf{G}_{d3mn}\,\Delta\mathbf{u}_M \end{bmatrix} , \tag{2.53$_1$}$$

$$\Delta\delta\mathbf{u}_{sn} = \begin{bmatrix} \delta\mathbf{u}_S{}^T\,\mathbf{G}_{d1sn}\,\Delta\mathbf{u}_S \\ \delta\mathbf{u}_S{}^T\,\mathbf{G}_{d2sn}\,\Delta\mathbf{u}_S \\ \delta\mathbf{u}_S{}^T\,\mathbf{G}_{d3sn}\,\Delta\mathbf{u}_S \end{bmatrix} . \tag{2.53$_2$}$$

These submatrices can be obtained from the partial differentiation of the component vectors in (2.50) with respect to the nodal displacements

$$\mathbf{G}_{djmn} = \frac{\partial \mathbf{G}_{jmn}}{\partial \mathbf{u}_M}, \qquad \mathbf{G}_{djsn} = \frac{\partial \mathbf{G}_{jsn}}{\partial \mathbf{u}_S}, \tag{2.54}$$

where $j = 1, 2$ or 3. However, carrying out these calculations in a direct way is very tedious and leads to extremely complex formulae. Hence, in the computer subroutine used to determine the tangent stiffness matrices (2.47) and (2.49), the partial derivatives (2.54) are evaluated numerically, using the finite difference method and small perturbations of the nodal displacements.

2.7 Numerical Examples

2.7.1 Introduction

In this section three numerical examples of the frictionless contact between beams with rectangular cross-sections are presented. All three are solved using the penalty method. Another examples of the frictional contact, which are also solved using the Lagrange multipliers method are given in Section 3.6. A numerical comparison of both methods is also presented therein.

It is well known, that the penalty method requires a proper selection of the value of the penalty parameter to avoid the ill-conditioning of the tangent stiffness matrix and, simultaneously, to fulfil the constraint conditions with an acceptable accuracy. In the presented examples the penalty parameter was chosen to keep the maximum penetration e_{tol} below 1% of the minimal dimension of the beams cross-sections.

The purpose of the presented examples is a qualitative analysis of contact and checking the correctness and the effectiveness of the proposed numerical algorithm. The physical solution of the problems is of a smaller importance, hence the data used in the examples are devoid of any physical units. It is understood, however, that any system of physically consistent units might be attached.

In these examples the loading is applied in the form of displacements imposed simultaneously in a given number of equal increments. The loading and the deformation process is measured in a fictitious time and a given moment in this time is determined by the parameter T, which ranges from 0 to 1.

Each of the examples is solved using the Newton-Raphson method. The convergence criterion uses a relative energy, which is calculated as a scalar product of the residual vector and the displacement increment vector related to such a product for the first iteration in a given increment. The convergence is assumed, if such a relative energy is smaller than 10^{-8}.

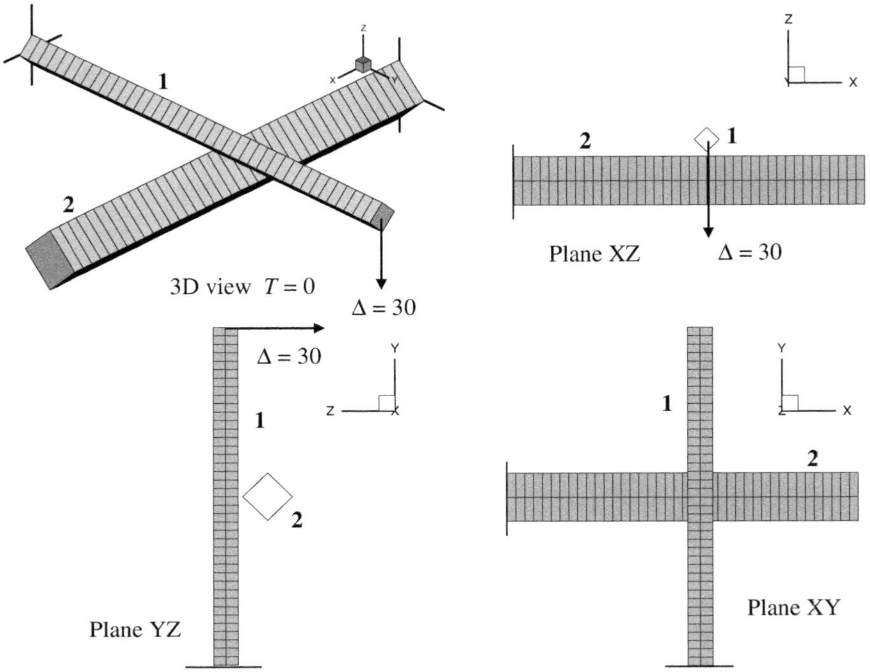

Fig. 2.8 Example 1 – initial configuration

2.7.2 Example 1

In this example contact between two cantilever beams shown in Fig. 2.8 is analyzed. Free end of beam **1** is subjected to displacement $\Delta = 30$, applied in 40 increments. The following data are used in the calculations: beam **1** – $E = 20000$, $\nu = 0{,}3$, cross-section dimensions $b_s = b_t = 5$, length 100; beam **2** – $E = 30000$, $\nu = 0.17$, cross-section dimensions $b_s = b_t = 10$, length 100; initial gap between the beams 1.393; penalty parameter $\varepsilon_N = 100000$.

In the first part of this example an influence of the beam element mesh on the results is investigated. In Table 2.1 values of dimensionless displacements of beams tips are presented depending on the number of elements. They are related to the respective displacements for the case, when each beam is divided into 16 elements – these values, considered as the exact ones are also given in the table. The values of maximal penetration are also included.

The obtained values of the displacements presented in the table indicate, that the division of the beams into 10 elements provides sufficiently accurate solution, for which the relative percentage errors calculated with respect to the division into 16 elements do not exceed 1%. Hence, in the following calculations such a discretisation into 10 elements per beam is adopted.

Table 2.1 Example 1 – dimensionless displacements of beam tips and penetration

Number of elements	Beam **1**			Beam **2**			Penetration
2, 2	0.951	0.927	1.000	0.883	1.063	0.965	0.98
4, 4	0.976	0.977	1.000	0.961	1.041	0.989	1.00
6, 6	0.985	0.990	1.000	0.971	1.025	0.995	1.00
8, 8	0.989	0.994	1.000	1.000	1.014	0.998	1.00
10, 10	0.993	0.998	1.000	1.000	1.007	0.999	1.00
12, 12	0.996	0.999	1.000	1.000	1.003	1.000	1.00
Reference solution							
16, 16	3.261	-8.260	-30.000	-0.103	-1.331	-4.147	0.0084

The penetration between the beams is almost insensitive to the finite element mesh. Its value, equal to 0.0084, represents only 0.17% of the smallest dimension of beams cross-sections and lies within the proposed range of tolerance.

In Fig. 2.9 beams configurations at four selected stages of the deformation process calculated with the discretisation into 10 elements per beam are presented. In this case, due to the considered values of the cross-sectional dimensions and the

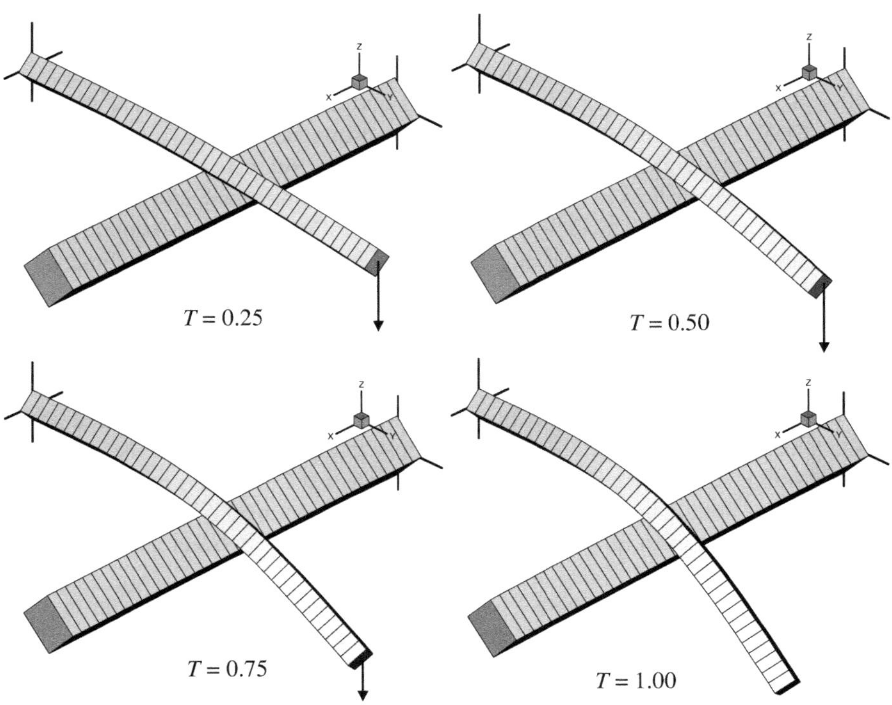

Fig. 2.9 Example 1 – deformed configurations of beams

Table 2.2 Example 1 – convergence results

Increment	T	Number of iterations	Penetration
10	0.25	3	0.00101
20	0.50	3	0.00338
30	0.75	4	0.00605
40	1.00	5	0.00840

elasticity moduli, beam **1** is the one, that deforms most. It is bent about beam **2** and also slides along it. In this example, from four possible pairs of the contacting edges, only one is active.

In Table 2.2 results of the convergence analysis of the iteration process are given. Till the 30th increment the convergence is very quick with just three iterations leading to the equilibrium state. Further on, the process is still convergent but not with quadratic rate and the number of required iterations increases to 5. It is due to the method of approximation of displacements of cross-section corners. The formula (2.16) is not accurate for large deformations, especially for large rotations, which occur at the last stages of the deformation process in this example.

2.7.3 Example 2

In this example a mutual contact between three beams presented in Fig. 2.10 is analysed. Two of these beams – cantilevers **1** and **2** are crossed in plane at the angle of 11.5°. Their free ends are subjected to displacements $\Delta_3 = \Delta_4 = 10$, which bring them into contact. The ends of beam **3** are subjected to displacements $\Delta_1 = \Delta_2 = 50$, which lead it to the contact with two cantilever beams. All the displacements are applied in 200 increments. The following data, identical for all beams, are taken to the calculations: $E = 20000$, $\nu = 0.3$, cross-section dimensions $b_s = b_t = 5$, length 100; initial gap between beams **1** and **2** 1.0; penalty parameter for contact between beams **1** and **2** $\varepsilon_{N1} = 10000$, penalty parameter for contact with beam **3** – $\varepsilon_{N2} = 100000$.

Beams configurations in four selected moments of the deformation process are presented in Fig. 2.11. Initially contact occurs only between beams **1** and **2** and it leads to increase of the angle in plane XY between the tangents to the beams at the contact point. After about 80 increments, i.e. for $T = 0.4$, beam **3** contacts simultaneously with both cantilever beams. Due to this contact beam **3** bends and at the same time causes even quicker increase of in-plane angle between axes of beams **1** and **2**.

In this example mutual contact between each pair of beams is present and the number of active edge pairs is different for each beam pair and varies in time. For the contact between beams **1** and **2** initially there is one edge pair active but during the process their number increases to three. On the other hand, beam **3** touches each of the cantilever beams at two edges.

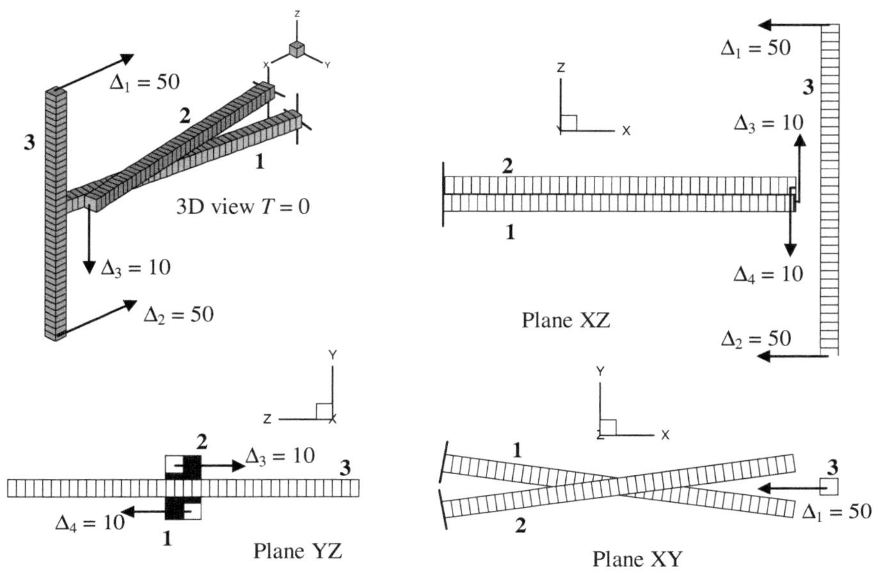

Fig. 2.10 Example 2 – initial configuration

Results of the convergence analysis are presented in Table 2.3. In this case with a relatively large number of increments it was enough to have just three iterations per increment to achieve the equilibrium state. Values of the penetration given in the same table do not exceed 0.3% of the minimal cross-sectional dimension.

Table 2.3 Example 2 – convergence results

Increment	T	No. of iterations	Penetration **1–2**	Penetration **1–3**
50	0.25	3	0.00619	No contact
100	0.50	3	0.01062	0.00047
150	0.75	3	0.01109	0.00166
200	1.00	3	0.01272	0.00396

2.7.4 Example 3

In this example a mutual contact between four identical beams forming a symmetric assembly shown in Fig. 2.12 is considered. Each beam is inclined to the plane XY by the angle of 11.5°. All free ends of beams are subjected to vertical displacements $\Delta = 20$, which bring all the beams to a simultaneous contact with their neighbours. Centre points of beams are partly fixed with the freedom of the rotation about axis X or Y, perpendicular to a given beam. All imposed displacements are applied in 100 increments. The following data, identical for all beams, are

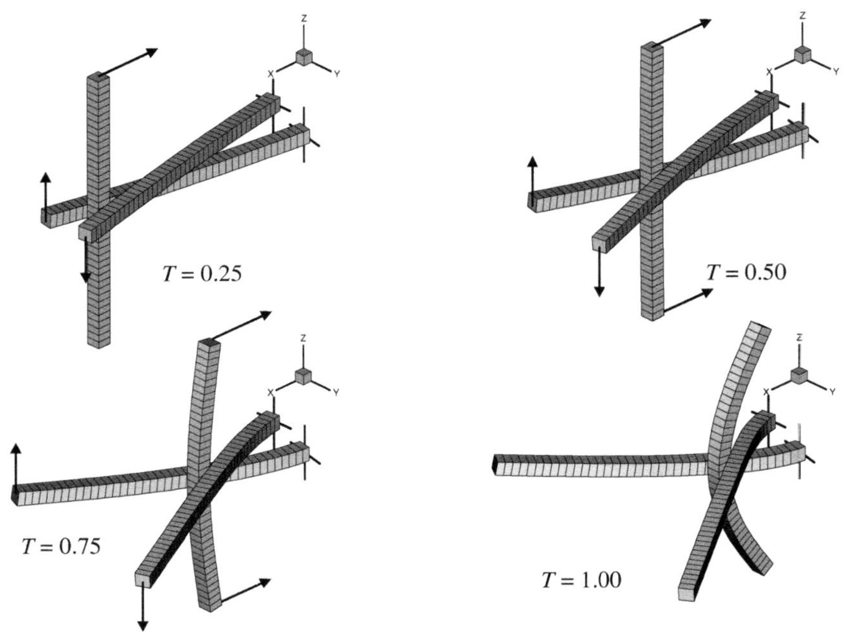

Fig. 2.11 Example 2 – deformed configurations of beams

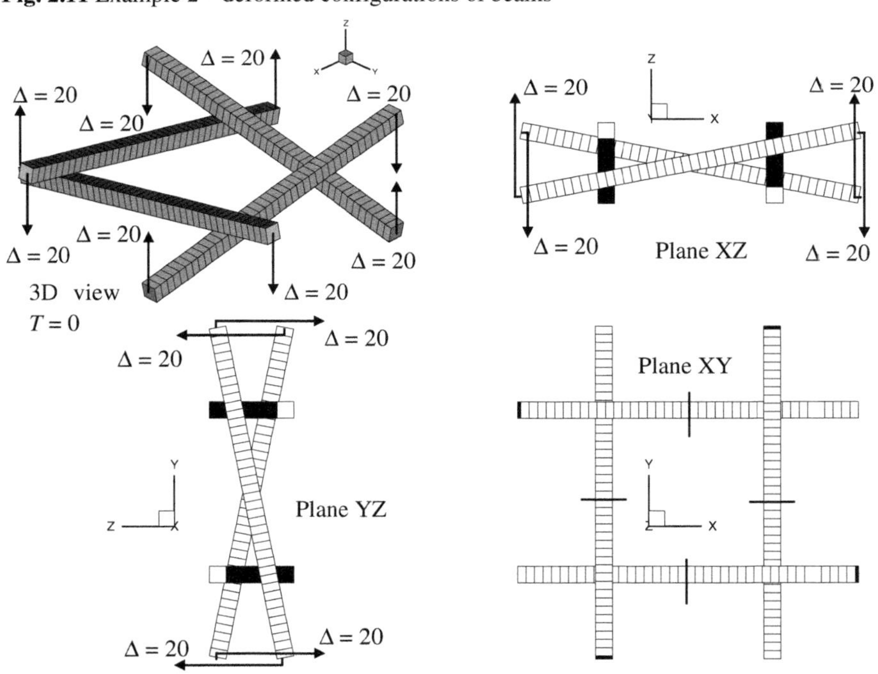

Fig. 2.12 Example 3 – initial configuration

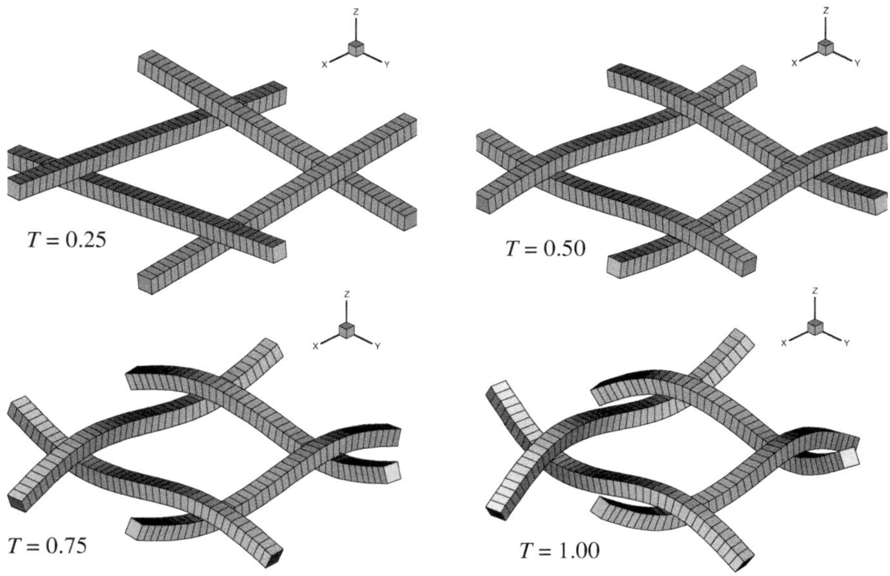

Fig. 2.13 Example 3 – deformed configurations of beams

Table 2.4 Example 3 – convergence results

Increment	T	Number of iterations	Penetration
25	0.25	3	0.00418
50	0.50	3	0.00962
75	0.75	4	0.03320
100	1.00	5	0.05069

taken to the calculations: $E = 20000$, $\nu = 0.3$, cross-section dimensions $b_s = b_t = 5$, length 100, penalty parameter $\varepsilon_N = 40000$.

Deformed configurations of the beams in the assembly for four selected moments of the deformation process are presented in Fig. 2.13. Due to the symmetry all four contact zones have the same characteristics. In the first stage only one pair of edges is active there and following the increase of displacement, penetration and rotations next pairs are switched on and at the end of the process all four edge pairs are active.

Results of the convergence analysis of the Newton-Raphson method are presented in Table 2.4. In all the increments 3 iterations were enough to achieve the equilibrium state. There are also the maximal penetration values presented. Their maximal value at the end of the process is 1.01% of the cross-section dimension – just over the assumed limit.

Chapter 3
Friction in Beam-to-Beam Contact

3.1 Friction Model

In the present analysis the Coulomb model of the dry friction with a constant friction coefficient μ between the friction force F_T and the normal force F_N in the contact point between beams is used

$$F_T = \mu F_N \,. \tag{3.1}$$

The advantage is taken of the analogy to the rigid-ideally plastic material suggested in the paper by Michałowski and Mróz (1978). This allows to distinguish between two friction states: the stick state, which is characterized by no relative displacement between the bodies, and the slip state, where the relative displacement in the form of sliding is present. The further considerations in this chapter concern one pair of edges or one pair of beam axes, like in the papers by Litewka and Wriggers (2002b, 2003). This can be easily extended to multiple pairs.

Two separate values of tangential displacements, g_{Tm} and g_{Ts}, are introduced. They correspond to two independent relative sliding movements on the beams m and s, respectively. According to the approach with the analogy to the plasticity, each of these displacements is divided into a stick part – elastic (superscript e) and a slip part – plastic (superscript p)

$$\begin{aligned} g_{Tm} &= g_{Tm}^{e} + g_{Tm}^{p} \,, \\ g_{Ts} &= g_{Ts}^{e} + g_{Ts}^{p} \,. \end{aligned} \tag{3.2}$$

Elastic parts are subjected to the constraint conditions (1.20). Their values are used to define the friction forces within the penalty method

$$\begin{aligned} F_{Tm} &= \varepsilon_{Tm} g_{Tm}^{e} = \varepsilon_{Tm}\left(g_{Tm} - g_{Tm}^{p}\right), \\ F_{Ts} &= \varepsilon_{Ts} g_{Ts}^{e} = \varepsilon_{Ts}\left(g_{Ts} - g_{Ts}^{p}\right). \end{aligned} \tag{3.3}$$

Such a formulation allows to introduce two different values of the penalty parameter. Hence, it is possible to distinguish between elastic properties of tangential contact for each beam.

P. Litewka: Finite Element Analysis of Beam-to-Beam Contact, LNACM 53, pp. 39–69.
springerlink.com

In the Lagrange multipliers method the friction forces are defined in the following way

$$F_{Tm} = \lambda_{Tm} \,,$$
$$F_{Ts} = \lambda_{Ts} \,. \tag{3.4}$$

Therefore, the introduction of friction in this method requires two additional unknowns – the Lagrange multipliers, λ_{Tm} and λ_{Ts}, for each pair of contacting beam edges or axes.

It is worth to note, that the friction forces F_{Tm} and F_{Ts} do not represent a force pair of action-reaction type. They are rather two independent quantities related to two independently considered displacements along two contacting beams. Each of this forces acts on both beams.

The plastic parts of the tangential displacements can be obtained from the sliding rule

$$\dot{g}^{p}_{Tm} = \dot{\gamma}\frac{\partial f_m}{\partial F_{Tm}}, \qquad \dot{g}^{p}_{Ts} = \dot{\gamma}\frac{\partial f_s}{\partial F_{Ts}}, \tag{3.5}$$

which corresponds to the non-associated flow rule in plasticity.

For the Coulomb friction law the counterpart of the plastic limit function – the sliding limit function, reads

$$f_m = \left|F_{Tm}\right| - \mu_m F_N \le 0 \,, \qquad f_s = \left|F_{Ts}\right| - \mu_s F_N \le 0 \,. \tag{3.6}$$

This formulation allows to introduce two different values of the friction coefficient for the beams – μ_m and μ_s. However, contrary to the similar situation with penalty parameters, such a distinction seems to have no physical justification and in further considerations only one friction coefficient for a beam pair is adopted

$$\mu = \mu_m = \mu_s \,. \tag{3.7}$$

The sign of the value of the sliding limit function (3.6) determines the current friction state. For non-positive values the stick state (elastic) is encountered, which is characterized by no relative displacements of the beams, while for positive values – the slip state (plastic) with the sliding occurs. In the latter case the solution requires the Euler procedure of return to the limit surface.

The sliding rules have to be integrated with respect to a fictitious time. To this end the incremental method can be applied. Here, the trial values of the elastic displacements in the current (new) step n are calculated

$$g^{et}_{Tmn} = g_{Tmn} - g^{p}_{Tmp} \,, \qquad g^{et}_{Tsn} = g_{Tsn} - g^{p}_{Tsp} \tag{3.8}$$

as differences between the current values of the total tangential displacements and the plastic parts of the tangential displacements from the previous step p. Then the trial values of the friction forces are found. Using the penalty method one gets

$$F_{Tmn}^{t} = \varepsilon_{Tm} g_{Tmn}^{et} ,\qquad F_{Tsn}^{t} = \varepsilon_{Ts} g_{Tsn}^{et} , \tag{3.9}$$

what in turn allows to obtain the trial values of the sliding function

$$f_{m}^{t} = \left| F_{Tmn}^{t} \right| - \mu F_{N} ,\qquad f_{s}^{t} = \left| F_{Tsn}^{t} \right| - \mu F_{N} . \tag{3.10}$$

If the sliding criterion (3.6) is fulfilled, then the contact is in the stick state. The real values of the friction forces are equal to their trial values

$$F_{Tmn} = F_{Tmn}^{t} ,\qquad F_{Tsn} = F_{Tsn}^{t} \tag{3.11}$$

and there is no update of the plastic parts of the tangential displacements

$$g_{Tmn}^{p} = g_{Tmp}^{p} ,\qquad g_{Tsn}^{p} = g_{Tsp}^{p} . \tag{3.12}$$

On the other hand, when the sliding criterion (3.6) is violated and any of the sliding functions (3.10) is positive, then the friction is in the slip state. A too large value of the friction force must be reduced to the maximum value corresponding to the sliding limit. In the case of the Coulomb friction law the Euler return procedure does not require an iterative approach and one obtains the closed-form solution for the friction forces

$$F_{Tmn} = \mu F_{N} \operatorname{sign}\left(g_{Tmn}^{et} \right),\qquad F_{Tsn} = \mu F_{N} \operatorname{sign}\left(g_{Tsn}^{et} \right) \tag{3.13}$$

and for the plastic parts of the tangential displacements

$$g_{Tmn}^{p} = g_{Tmp}^{p} + \frac{1}{\varepsilon_{Tm}} \left[F_{Tmn}^{t} - F_{Tmn} \operatorname{sign}\left(g_{Tmn}^{et} \right) \right],\qquad g_{Tsn}^{p} = g_{Tsp}^{p} + \frac{1}{\varepsilon_{Ts}} \left[F_{Tsn}^{t} - F_{Tsn} \operatorname{sign}\left(g_{Tsn}^{et} \right) \right]. \tag{3.14}$$

The orientation of the displacements is controlled by the sign of the value of its trial elastic part.

When the Lagrange multipliers method is used, then the constraint condition (1.20) applied to the elastic part of the tangential displacement is fulfilled exactly. Hence, the entire value of the tangential displacement is equal to its plastic part – it is non-zero only in the slip state. The trial values of the friction force are expressed by the Lagrange multipliers

$$\begin{aligned} F_{Tmn}^{t} &= \lambda_{Tmn} , \\ F_{Tsn}^{t} &= \lambda_{Tsn} . \end{aligned} \tag{3.15}$$

Then the value of the sliding function (3.10) is calculated and the sliding criterion is checked. In the case of the stick state the real value of the friction force is equal to the trial force, like in (3.11) and with the vanishing value of the elastic part of the tangential displacement its total value remains unchanged.

In the slip state the value of the friction force is returned to the sliding limit surface, what leads to the following relations between the Lagrange multipliers

$$\begin{aligned} \left|\lambda_{Tmn}\right| &= \mu\lambda_N , \\ \left|\lambda_{Tsn}\right| &= \mu\lambda_N . \end{aligned} \tag{3.16}$$

The plastic part of the tangential displacement represents also its total value.

It is possible, that during the deformation process friction can switch from slip to stick state. In the penalty method this change does not pose any additional problems. In the Lagrange multipliers method such a situation must be recognized, when the absolute value of the tangential displacement starts to decrease.

In the above presented considerations it is suggested, that the sliding criterion is checked independently for two friction forces acting along edges of beams m and s. Hence, in this approach a possibility exists, that in one contact point two different friction states – stick and slip are present at the same time.

To eliminate such a problematic case a different approach might be used. The forces F_{Tmn} and F_{Tsn} can be considered as two components of one resultant friction force for each contact point. In such a situation the sliding criterion is checked only once in the form

$$f = \left\|\mathbf{F}_T\right\| - \mu F_N \leq 0 , \tag{3.17}$$

where the value of the resultant friction force in contact can be calculated from

$$\left\|\mathbf{F}_T{}^t\right\| = F_T{}^t = \left\|F_{Tmn}{}^t\,\mathbf{t}_m + F_{Tsn}{}^t\,\mathbf{t}_s\right\| \tag{3.18}$$

The unit tangent vectors $\mathbf{t}_m$ and $\mathbf{t}_s$ required in (3.18) are defined in a detailed way in Section 3.2 by Eq. (3.30).

In the stick state all the formulae given previously are valid, while in the slip state the resultant friction force is limited by the maximal value μF_N, and its components in the directions of the tangent vectors $\mathbf{t}_m$ and $\mathbf{t}_s$ can be determined from the following relations

$$F_{Tmn}{}^t = \mu p_m F_N ,$$
$$F_{Tsn}{}^t = \mu p_s F_N , \tag{3.19}$$

where the proportion parameters are introduced

$$p_m = \frac{F_{Tmn}{}^t}{F_T{}^t} ,$$
$$p_s = \frac{F_{Tsn}{}^t}{F_T{}^t} . \tag{3.20}$$

In this approach for the Lagrange multipliers method formulae (3.16) must be replaced with the following

$$|\lambda_{Tmn}| = \mu p_m \lambda_N ,$$
$$|\lambda_{Tsn}| = \mu p_s \lambda_N \tag{3.21}$$

to limit the friction force components.

3.2 Kinematic Variables for Friction

The tangential displacement g_T, which is a basic kinematic variable can be calculated using various methods. Generally speaking, this variable represents the distance, through which the contact point traveled on the body surface. This value depends on the path, not just on the initial and final position, and the integration of the velocity with respect to time is necessary. To this end, the approximate incremental procedure can be used, which leads to summation of the increments d_{gTmn} and d_{gTsn}

$$g_{Tmn} = g_{Tmp} + d_{gTmn} ,$$
$$g_{Tsn} = g_{Tsp} + d_{gTsn} . \tag{3.22}$$

Usually, the location of contact points is determined using directly the local co-ordinates. But in the case of contact between beams with rectangular cross-sections where the local co-ordinates are related to the beam axis and the contact points lie on the edges, it is more convenient to use the position vectors. Such an approach was suggested in the paper by Agelet de Saracibar (1997). Besides, independently of the cross-section shape, such a formulation allows for an easier treatment of the case, when the contact point moves from one finite element to another.

The displacement increment can be calculated in the previous or in the current configuration, see Fig. 3.1. In the first case on can write down

$$\begin{aligned} \left| d_{gTmn} \right| &= \left\| \mathbf{x}_{mnb} - \mathbf{x}_{mp} \right\|, \\ \left| d_{gTsn} \right| &= \left\| \mathbf{x}_{snb} - \mathbf{x}_{sp} \right\|, \end{aligned} \tag{3.23}$$

where the position vectors of the contact points in the previous configuration, $\mathbf{x}_{mp}$ and $\mathbf{x}_{sp}$, and the position vectors of the current contact points, $\mathbf{x}_{mnb}$ and $\mathbf{x}_{snb}$, mapped on the previous configuration are used. Two latter vectors are calculated using the current values of the local co-ordinates of the contact points ξ_{mn} and ξ_{sn}.

On the other hand, if the increments are calculated in the current configuration the following relations can be used

$$\begin{aligned} \left| d_{gTmn} \right| &= \left\| \mathbf{x}_{mn} - \mathbf{x}_{mpf} \right\|, \\ \left| d_{gTsn} \right| &= \left\| \mathbf{x}_{sn} - \mathbf{x}_{spf} \right\|. \end{aligned} \tag{3.24}$$

Here the position vectors of the current contact points, $\mathbf{x}_{mn}$ and $\mathbf{x}_{sn}$, and the position vectors of the previous contact points $\mathbf{x}_{mpf}$ and $\mathbf{x}_{spf}$ mapped on the current configuration (see Fig. 3.1) are involved. The latter vectors are obtained using the local co-ordinates for the previous position of the contact points ξ_{mp} and ξ_{sp}.

In the both presented methods curvilinear paths, along which the contact points travel at a given load increment are approximated by straight lines. Such an approximation can be regarded as justified in the case of the beam-to-beam contact when small strains are assumed (Litewka and Wriggers 2002b). In the following the second approach is used, where the increment of the tangential displacement is calculated in the current configuration following Eq. (3.24).

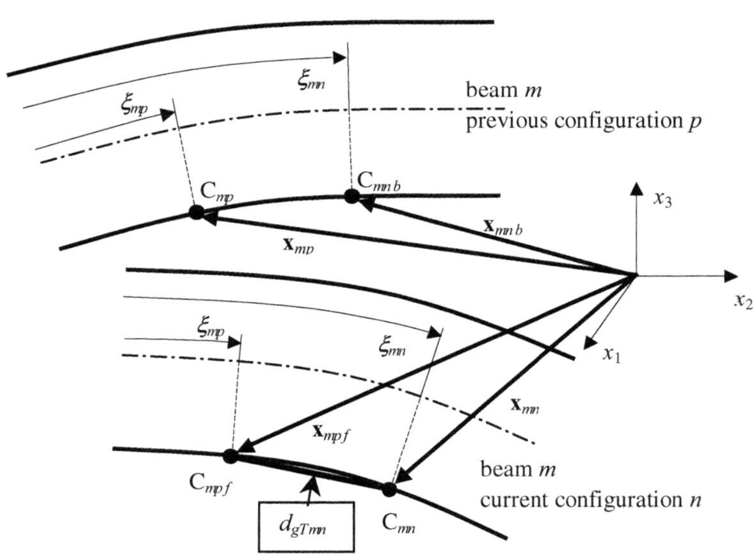

Fig. 3.1 Increment of tangential displacement

The absolute values of these increments lack the sign, which must be added to distinguish between two possibilities of sliding in both orientations along the beam. If the sliding takes place within just one element, the local co-ordinates can be used

$$\begin{aligned} s_m &= \operatorname{sign}\left(\xi_{mn} - \xi_{mp}\right), \\ s_s &= \operatorname{sign}\left(\xi_{sn} - \xi_{sp}\right). \end{aligned} \tag{3.25}$$

On the other hand, if the contact point moves between several elements a different definition should be used. For instance, a sequential numbering of beam elements can be adopted and then s_m and s_s could be determined as the signs of the differences of the element numbers.

However, there is a problem with formulae (3.25), when the Lagrange multipliers method is used. With the exact fulfillment of the constraint equation in the stick state, the local co-ordinates of the contact points in the subsequent increments are equal. In the numerical analysis this would provide chaotic changes of values of s_m and s_s. To avoid this problem the following relations should be used

$$\begin{aligned} s_m &= \operatorname{sign}\left(\lambda_{Tm}\right), \\ s_s &= \operatorname{sign}\left(\lambda_{Ts}\right). \end{aligned} \tag{3.26}$$

Finally, the current values of the tangential displacements can be expressed as

$$g_{Tmn} = g_{Tmp} + s_m \left\| \mathbf{x}_{mn} - \mathbf{x}_{mpf} \right\|, \tag{3.27$_1$}$$

$$g_{Tsn} = g_{Tsp} + s_s \left\| \mathbf{x}_{sn} - \mathbf{x}_{spf} \right\|. \tag{3.27$_2$}$$

Now the variations, the linearisations and the linearisations of the variations of these variables must be derived. It is worth to note, that the local co-ordinates of the previous contact points are fixed at the current increment. Hence, their variations and linearisations vanish and the variations and the linearizations of the position vectors of the previous contact points have somewhat simpler form

$$\begin{aligned} \delta\mathbf{x}_{mpf} &= \delta\mathbf{u}_{mp}, \\ \delta\mathbf{x}_{spf} &= \delta\mathbf{u}_{sp}, \\ \Delta\mathbf{x}_{mpf} &= \Delta\mathbf{u}_{mp}, \\ \Delta\mathbf{x}_{spf} &= \Delta\mathbf{u}_{sp}, \end{aligned} \tag{3.28}$$

contrary to their counterparts related to the position vectors of the current contact points, which were given by Eq. (2.24). After some mathematical operations the required kinematic variables can be determined (Zavarise and Wriggers 2000, Litewka and Wriggers 2002b)

$$\begin{aligned}
\delta g_{Tmn} &= s_m \mathbf{t}_m \circ \left(\mathbf{x}_{mn,m}\delta\xi_{mn} + \delta\mathbf{u}_{mn} - \delta\mathbf{u}_{mp}\right),\\
\delta g_{Tsn} &= s_s \mathbf{t}_s \circ \left(\mathbf{x}_{sn,s}\delta\xi_{sn} + \delta\mathbf{u}_{sn} - \delta\mathbf{u}_{sp}\right),\\
\Delta g_{Tmn} &= s_m \mathbf{t}_m \circ \left(\mathbf{x}_{mn,m}\Delta\xi_{mn} + \Delta\mathbf{u}_{mn} - \Delta\mathbf{u}_{mp}\right),\\
\Delta g_{Tsn} &= s_s \mathbf{t}_s \circ \left(\mathbf{x}_{sn,s}\Delta\xi_{sn} + \Delta\mathbf{u}_{sn} - \Delta\mathbf{u}_{sp}\right),\\
\Delta\delta g_{Tmn} &= \frac{s_m}{d_{gTmn}}\left(\mathbf{x}_{mn,m}\delta\xi_{mn} + \delta\mathbf{u}_{mn} - \delta\mathbf{u}_{mp}\right)\left(1 - \mathbf{t}_m \otimes \mathbf{t}_m\right)\\
&\quad \left(\mathbf{x}_{mn,m}\Delta\xi_{mn} + \Delta\mathbf{u}_{mn} - \Delta\mathbf{u}_{mp}\right) + s_m \mathbf{t}_m \circ \left(\mathbf{x}_{mn,mm}\delta\xi_{mn}\Delta\xi_{mn} + \right.\\
&\quad \left. + \Delta\mathbf{u}_{mn,m}\delta\xi_{mn} + \delta\mathbf{u}_{mn,m}\Delta\xi_{mn} + \mathbf{x}_{mn,m}\Delta\delta\xi_{mn} + \Delta\delta\mathbf{u}_{mn} - \Delta\delta\mathbf{u}_{mp}\right),\\
\Delta\delta g_{Tsn} &= \frac{s_s}{d_{gTsn}}\left(\mathbf{x}_{sn,s}\delta\xi_{sn} + \delta\mathbf{u}_{sn} - \delta\mathbf{u}_{sp}\right)\left(1 - \mathbf{t}_s \otimes \mathbf{t}_s\right)\\
&\quad \left(\mathbf{x}_{sn,s}\Delta\xi_{sn} + \Delta\mathbf{u}_{sn} - \Delta\mathbf{u}_{sp}\right) + s_s \mathbf{t}_s \circ \left(\mathbf{x}_{sn,ss}\delta\xi_{sn}\Delta\xi_{sn} + \right.\\
&\quad \left. + \Delta\mathbf{u}_{sn,s}\delta\xi_{sn} + \delta\mathbf{u}_{sn,s}\Delta\xi_{sn} + \mathbf{x}_{sn,s}\Delta\delta\xi_{sn} + \Delta\delta\mathbf{u}_{sn} - \Delta\delta\mathbf{u}_{sp}\right).
\end{aligned} \tag{3.29}$$

In the above formulae two unit tangent vectors are introduced. They can be obtained from the following relations

$$\mathbf{t}_m = \frac{\mathbf{x}_{mn} - \mathbf{x}_{mpf}}{d_{gTmn}} = \left(t_{m1}, t_{m2}, t_{m3}\right)^T, \tag{3.30$_1$}$$

$$\mathbf{t}_s = \frac{\mathbf{x}_{sn} - \mathbf{x}_{spf}}{d_{gTsn}} = \left(t_{s1}, t_{s2}, t_{s3}\right)^T. \tag{3.30$_2$}$$

If the Lagrange multipliers method is used and the constraints are fulfilled exactly, the calculation of the variables $\Delta\delta g_{Tmn}$ $(3.29)_5$ and $\Delta\delta g_{Tsn}$ $(3.29)_6$, containing d_{gTmn} and d_{gTsn}, respectively, in the denominators leads to a numerical problem of division by a very small number. To avoid this a different definition of the tangent vectors can be used, where the derivatives of the position vectors of the current contact points with respect to the local co-ordinates are involved

$$\begin{aligned}
\mathbf{t}_m &= \frac{\mathbf{x}_{mn,m}}{\left\|\mathbf{x}_{mn,m}\right\|},\\
\mathbf{t}_m &= \frac{\mathbf{x}_{sn,s}}{\left\|\mathbf{x}_{sn,s}\right\|}.
\end{aligned} \tag{3.31}$$

After such a modification the denominators in Eqs. $(3.29)_{5,6}$ contain non-zero lengths of the derivative vectors, which are present in the denominators of (3.31).

Additional attention must be drawn to the linearisations of the variations of the local-co-ordinates of the current contact points, $\Delta\delta\xi_{mn}$ and $\Delta\delta\xi_{sn}$, which are also present in Eqs. $(3.29)_5$ and $(3.29)_6$. Their derivation requires very tedious calculations, which are summarised in Appendix 2.

3.3 Weak Form Components due to Friction

Each of the components resulting from the presence of friction in the active constraints, which must be included in the functional (1.21) or (1.22), can be expressed in the following form

$$\delta\Pi_T = \delta\Pi_{Tm} + \delta\Pi_{Ts} = F_{Tmn}\delta g_{Tmn} + F_{Tsn}\delta g_{Tsn} . \tag{3.32}$$

It should be pointed out, that in the analysis of the beam-to-beam contact two separate terms are introduced. They correspond to the relative displacements of contact point along each of the contacting beam edges or axes. The related friction forces can be considered as independent (Zavarise and Wriggers 2000, Litewka and Wriggers 2002b) or as components of one resultant force (Litewka 2006). In the following the frictional beam-to-beam contact formulation for these both possibilities is presented.

Solution of the nonlinear problem in hand requires calculation of the linearisation of the expression (3.32)

$$\begin{aligned}\Delta\delta\Pi_T &= \Delta\delta\Pi_{Tm} + \Delta\delta\Pi_{Ts} = \\ &= \Delta F_{Tmn}\delta g_{Tmn} + F_{Tmn}\Delta\delta g_{Tmn} + \Delta F_{Tsn}\delta g_{Tsn} + F_{Tsn}\Delta\delta g_{Tsn} .\end{aligned} \tag{3.33}$$

The value of the friction force depends on the method used to satisfy the constraint conditions, on the current friction state, i.e. stick or slip and on the way to treat the friction forces along each of the contacting beams.

With the penalty method (superscript ε) in the stick state (superscript e) one gets the following relations

$$\begin{aligned}\delta\Pi_{Tm}^{e\varepsilon} &= \varepsilon_{Tm}\, g_{Tmn}^{e}\, \delta g_{Tmn} ,\\ \delta\Pi_{Ts}^{e\varepsilon} &= \varepsilon_{Ts}\, g_{Tsn}^{e}\, \delta g_{Tsn} ;\end{aligned} \tag{3.34}$$

$$\begin{aligned}\Delta\delta\Pi_{Tm}^{e\varepsilon} &= \varepsilon_{Tm}\, \Delta g_{Tmn}\, \delta g_{Tmn} + \varepsilon_{Tm}\, g_{Tmn}^{e}\, \Delta\delta g_{Tmn} ,\\ \Delta\delta\Pi_{Ts}^{e\varepsilon} &= \varepsilon_{Ts}\, \Delta g_{Tsn}\, \delta g_{Tsn} + \varepsilon_{Ts}\, g_{Tsn}^{e}\, \Delta\delta g_{Tsn} .\end{aligned} \tag{3.35}$$

In the slip state the formulae depend on the way to treat the friction forces and the method of checking the sliding criterion. In the case of two independent variables and the criterion (3.10) the weak form terms and their linearisations read

$$\begin{aligned}\delta\Pi_{Tm}^{p\varepsilon} &= \mu\varepsilon_N g_N\, \delta g_{Tmn} ,\\ \delta\Pi_{Ts}^{p\varepsilon} &= \mu\varepsilon_N\, g_N\, \delta g_{Tsn} ;\end{aligned} \tag{3.36}$$

$$\begin{aligned}\Delta\delta\Pi_{Tm}^{p\varepsilon} &= \mu\varepsilon_N\, \Delta g_N\, \delta g_{Tmn} + \mu\varepsilon_N\, g_N\, \Delta\delta g_{Tmn} ,\\ \Delta\delta\Pi_{Ts}^{p\varepsilon} &= \mu\varepsilon_N\, \Delta g_N\, \delta g_{Tsn} + \mu\varepsilon_N\, g_N\, \Delta\delta g_{Tsn} .\end{aligned} \tag{3.37}$$

On the other hand, if the resultant friction force is calculated from two component friction forces, then the sliding criterion has the form (3.17). Hence, the friction forces are calculated from Eq. (3.19) and one must take into account dependence of the proportion parameters (3.20) on displacements. That is why their linearisations are non-zero and must be taken into account in the derivation. In this case one gets

$$\begin{aligned} \delta\Pi_{Tm}^{p\varepsilon} &= \mu\varepsilon_N\ p_m\ s_m\ g_N\ \delta g_{Tmn}\ , \\ \delta\Pi_{Ts}^{p\varepsilon} &= \mu\varepsilon_N\ p_s\ s_s\ g_N\ \delta g_{Tsn}\ ; \end{aligned} \tag{3.38}$$

$$\begin{aligned} \Delta\delta\Pi_{Tm}^{p\varepsilon} &= \mu\varepsilon_N\ p_m\ s_m\ \Delta g_N\ \delta g_{Tmn} + \mu\varepsilon_N\ p_m\ s_m\ g_N\ \Delta\delta g_{Tmn} + \\ &\quad + \mu\varepsilon_N\ s_m\ g_N\ \Delta p_m\ \delta g_{Tmn}\ , \\ \Delta\delta\Pi_{Ts}^{p\varepsilon} &= \mu\varepsilon_N\ p_s\ s_s\ \Delta g_N\ \delta g_{Tsn} + \mu\varepsilon_N\ p_s\ s_s\ g_N\ \Delta\delta g_{Tsn} + \\ &\quad + \mu\varepsilon_N\ s_s\ g_N\ \Delta p_s\ \delta g_{Tsn} \end{aligned} \tag{3.39}$$

In this formulation the orientation of sliding is included in the sign of proportion parameters and also in the variation of the tangential displacements. Their product is always positive, so it is necessary to introduce additionally values s_m and s_s in Eqs. (3.38) and (3.39).

If the Lagrange multipliers method is used in the analysis (superscript λ), then in the stick state the following expressions are valid

$$\begin{aligned} \delta\Pi_{Tm}^{e\lambda} &= \lambda_{Tm}\delta g_{Tmn} + \delta\lambda_{Tm} g_{Tmn}\ , \\ \delta\Pi_{Ts}^{e\lambda} &= \lambda_{Ts}\delta g_{Tsn} + \delta\lambda_{Ts} g_{Tsn}\ ; \end{aligned} \tag{3.40}$$

$$\begin{aligned} \Delta\delta\Pi_{Tm}^{e\lambda} &= \Delta\lambda_{Tm}\delta g_{Tmn} + \delta\lambda_{Tm}\Delta g_{Tmn} + \lambda_{Tm}\Delta\delta g_{Tmn}\ , \\ \Delta\delta\Pi_{Ts}^{e\lambda} &= \Delta\lambda_{Ts}\delta g_{Tsn} + \delta\lambda_{Ts}\Delta g_{Tsn} + \lambda_{Ts}\Delta\delta g_{Tsn}\ . \end{aligned} \tag{3.41}$$

In the slip state with the independent check of the sliding criterion for two components of the friction force one gets

$$\begin{aligned} \delta\Pi_{Tm}^{p\lambda} &= \mu\lambda_N\delta g_{Tmn}\ , \\ \delta\Pi_{Ts}^{p\lambda} &= \mu\lambda_N\delta g_{Tsn}\ ; \end{aligned} \tag{3.42}$$

$$\begin{aligned} \Delta\delta\Pi_{Tm}^{p\lambda} &= \mu\Delta\lambda_N\delta g_{Tmn} + \mu\lambda_N\Delta\delta g_{Tmn}\ , \\ \Delta\delta\Pi_{Ts}^{p\lambda} &= \mu\Delta\lambda_N\delta g_{Tsn} + \mu\lambda_N\Delta\delta g_{Tsn}\ , \end{aligned} \tag{3.43}$$

while with the resultant friction force and a single check of the sliding criterion –

$$\begin{aligned} \delta\Pi_{Tm}^{p\lambda} &= \mu\lambda_N p_m s_m\delta g_{Tmn}\ , \\ \delta\Pi_{Ts}^{p\lambda} &= \mu\lambda_N p_s s_s\delta g_{Tsn}\ ; \end{aligned} \tag{3.44}$$

$$
\begin{aligned}
\Delta\delta\Pi_{Tm}^{p\lambda} &= \mu p_m s_m \Delta\lambda_N \delta g_{Tmn} + \mu p_m s_m \lambda_N \Delta\delta g_{Tmn} + \mu s_m \lambda_N \Delta p_m \delta g_{Tmn} , \\
\Delta\delta\Pi_{Ts}^{p\lambda} &= \mu p_s s_s \Delta\lambda_N \delta g_{Tsn} + \mu p_s s_s \lambda_N \Delta\delta g_{Tsn} + \mu s_s \lambda_N \Delta p_s \delta g_{Tsn} .
\end{aligned} \tag{3.45}
$$

In the above equations, similarly as in their counterparts for the penalty method (3.38) and (3.39), the sliding orientation is controlled by the values s_m and s_s.

It is worth to note, that in the Lagrange multipliers method the values of the multipliers corresponding to friction, λ_{Tm} and λ_{Ts}, can be negative. Such a situation takes place, when the sliding has the orientation opposite to the local co-ordinate. Besides, in the slip state the set of equations must be completed with the components, which would ensure fulfilment of the conditions (3.16) or (3.21), limiting the values of the friction forces. These terms are introduced in Section 3.5.

Using the resultant friction force in the sliding criterion requires calculation of the linearisation of the proportion coefficients p_m and p_s. Carrying out the appropriate mathematical operations to (3.20) leads to the following

$$
\begin{aligned}
\Delta p_m &= \left[\frac{\varepsilon_{Tm}}{F_T^{\,t}} - \frac{p_m\,\varepsilon_{Tm}}{\left(F_T^{\,t}\right)^2} \mathbf{F}_T^{\,t} \circ \mathbf{t}_m \right] \Delta g_{Tmn} - \frac{p_m\,\varepsilon_{Ts}}{\left(F_T^{\,t}\right)^2} \mathbf{F}_T^{\,t} \circ \mathbf{t}_s \,\Delta g_{Tsn} \\
&- \frac{p_m\,\varepsilon_{Tm}\, g_{Tmn}^{e}}{\left(F_T^{\,t}\right)^2 \left\| \mathbf{x}_{mn} - \mathbf{x}_{mpf} \right\|} \mathbf{F}_T^{\,t} \left(\mathbf{1} - \mathbf{t}_m \otimes \mathbf{t}_m \right) \left(\mathbf{x}_{mn,m} \Delta\xi_{mn} + \Delta\mathbf{u}_{mn} - \Delta\mathbf{u}_{mp} \right) \\
&- \frac{p_m\,\varepsilon_{Ts}\, g_{Tsn}^{e}}{\left(F_T^{\,t}\right)^2 \left\| \mathbf{x}_{sn} - \mathbf{x}_{spf} \right\|} \mathbf{F}_T^{\,t} \left(\mathbf{1} - \mathbf{t}_s \otimes \mathbf{t}_s \right) \left(\mathbf{x}_{sn,s} \Delta\xi_{sn} + \Delta\mathbf{u}_{sn} - \Delta\mathbf{u}_{sp} \right), \\
\Delta p_s &= \left[\frac{\varepsilon_{Ts}}{F_T^{\,t}} - \frac{p_s\,\varepsilon_{Ts}}{\left(F_T^{\,t}\right)^2} \mathbf{F}_T^{\,t} \circ \mathbf{t}_s \right] \Delta g_{Tsn} - \frac{p_s\,\varepsilon_{Tm}}{\left(F_T^{\,t}\right)^2} \mathbf{F}_T^{\,t} \circ \mathbf{t}_m \,\Delta g_{Tmn} \\
&- \frac{p_s\,\varepsilon_{Tm}\, g_{Tmn}^{e}}{\left(F_T^{\,t}\right)^2 \left\| \mathbf{x}_{mn} - \mathbf{x}_{mpf} \right\|} \mathbf{F}_T^{\,t} \left(\mathbf{1} - \mathbf{t}_m \otimes \mathbf{t}_m \right) \left(\mathbf{x}_{mn,m} \Delta\xi_{mn} + \Delta\mathbf{u}_{mn} - \Delta\mathbf{u}_{mp} \right) \\
&- \frac{p_s\,\varepsilon_{Ts}\, g_{Tsn}^{e}}{\left(F_T^{\,t}\right)^2 \left\| \mathbf{x}_{sn} - \mathbf{x}_{spf} \right\|} \mathbf{F}_T^{\,t} \left(\mathbf{1} - \mathbf{t}_s \otimes \mathbf{t}_s \right) \left(\mathbf{x}_{sn,s} \Delta\xi_{sn} + \Delta\mathbf{u}_{sn} - \Delta\mathbf{u}_{sp} \right).
\end{aligned} \tag{3.46}
$$

With this in hand all the necessary kinematic variables present in the weak form for frictional contact and its linearisation are found.

3.4 Discretization of Friction Terms Present in Weak Form

Some of the necessary variables have already been discretised in Sections 2.5 and 2.6. The lacking terms are the kinematic variables related to the previous position of the contact point – the variations, the linearisations and the linearisations of the variations. They can be obtained in the same way as their counterparts for the current contact points (2.34), (2.35) and (2.53). This leads to the following relations

$$\Delta\mathbf{u}_{mp} = \mathbf{G}_{mp}\Delta\mathbf{u}_M \,, \quad \Delta\mathbf{u}_{sp} = \mathbf{G}_{sp}\Delta\mathbf{u}_S \,, \tag{3.47}$$

$$\delta\mathbf{u}_{mp} = \mathbf{G}_{mp}\delta\mathbf{u}_M \,, \quad \delta\mathbf{u}_{sp} = \mathbf{G}_{sp}\delta\mathbf{u}_S \,. \tag{3.48}$$

The matrices $\mathbf{G}_{mp}$ and $\mathbf{G}_{sp}$ are calculated in the similar way as $\mathbf{G}_{mn}$ and $\mathbf{G}_{sn}$ using the local co-ordinates corresponding to the previous locations of the contact points – ξ_{mp} and ξ_{sp}. Each of these matrices can be expressed by three 12-element vectors

$$\mathbf{G}_{mp} = \left[\mathbf{G}_{1mp}, \mathbf{G}_{2mp}, \mathbf{G}_{3mp}\right]^T , \quad \mathbf{G}_{sp} = \left[\mathbf{G}_{1sp}, \mathbf{G}_{2sp}, \mathbf{G}_{3sp}\right]^T . \tag{3.49}$$

Differentiation of these vectors with respect to the nodal displacements, where, similarly as in the case of vectors (2.50) it is more convenient to use the method of small perturbations and the finite difference scheme, yields

$$\mathbf{G}_{djmp} = \frac{\partial \mathbf{G}_{jmp}}{\partial \mathbf{u}_M} , \quad \mathbf{G}_{djsp} = \frac{\partial \mathbf{G}_{jsp}}{\partial \mathbf{u}_S} , \tag{3.50}$$

where $j = 1, 2$ or 3. Then the linearisations of the variations of the displacements in the previous contact points can be expressed as

$$\Delta\delta\mathbf{u}_{mp} = \begin{bmatrix} \delta\mathbf{u}_M{}^T\, \mathbf{G}_{d1mp}\, \Delta\mathbf{u}_M \\ \delta\mathbf{u}_M{}^T\, \mathbf{G}_{d2mp}\, \Delta\mathbf{u}_M \\ \delta\mathbf{u}_M{}^T\, \mathbf{G}_{d3mp}\, \Delta\mathbf{u}_M \end{bmatrix} , \quad \Delta\delta\mathbf{u}_{sp} = \begin{bmatrix} \delta\mathbf{u}_S{}^T\, \mathbf{G}_{d1sp}\, \Delta\mathbf{u}_S \\ \delta\mathbf{u}_S{}^T\, \mathbf{G}_{d2sp}\, \Delta\mathbf{u}_S \\ \delta\mathbf{u}_S{}^T\, \mathbf{G}_{d3sp}\, \Delta\mathbf{u}_S \end{bmatrix} . \tag{3.51}$$

Discretisation must also be carried out for the linearisations of the proportion parameters (3.46). To simplify the notation the following matrices are introduced

$$\mathbf{S}_m = \mathbf{x}_{mn,m}\,\mathbf{F}_m + \left[\left(\mathbf{G}_{mn} - \mathbf{G}_{mp}\right) \;\; \mathbf{0}\right], \quad \mathbf{S}_s = \mathbf{x}_{sn,s}\,\mathbf{F}_s + \left[\mathbf{0} \;\; \left(\mathbf{G}_{sn} - \mathbf{G}_{sp}\right)\right], \tag{3.52}$$

where the split of the matrix $\mathbf{F}$ into the component vectors, according to Eq. (A2.3) is used. After these simplifications the following matrix relations are obtained

$$
\begin{aligned}
\Delta p_m = & \left\{ \left[\frac{\varepsilon_{Tm}}{F_T{}^t} - \frac{p_m \varepsilon_{Tm}}{\left(F_T{}^t\right)^2} \mathbf{F}_T{}^t \circ \mathbf{t}_m \right] \mathbf{t}_m{}^T \mathbf{S}_m \right. \\
& - \frac{p_m \varepsilon_{Tm} g_{Tmn}^e}{\left(F_T{}^t\right)^2 \left\| \mathbf{x}_{mn} - \mathbf{x}_{mpf} \right\|} \mathbf{F}_T{}^t \left(\mathbf{1} - \mathbf{t}_m \otimes \mathbf{t}_m\right) \mathbf{S}_m - \frac{p_m \varepsilon_{Ts}}{\left(F_T{}^t\right)^2} \left(\mathbf{F}_T{}^t \circ \mathbf{t}_s\right) \mathbf{t}_s{}^T \mathbf{S}_s \\
& \left. - \frac{p_m \varepsilon_{Ts} g_{Tsn}^e}{\left(F_T{}^t\right)^2 \left\| \mathbf{x}_{sn} - \mathbf{x}_{spf} \right\|} \mathbf{F}_T{}^t \left(\mathbf{1} - \mathbf{t}_s \otimes \mathbf{t}_s\right) \mathbf{S}_s \right\} \begin{bmatrix} \Delta \mathbf{u}_M \\ \Delta \mathbf{u}_S \end{bmatrix}, \\
\Delta p_s = & \left\{ \left[\frac{\varepsilon_{Ts}}{F_T{}^t} - \frac{p_s \varepsilon_{Ts}}{\left(F_T{}^t\right)^2} \mathbf{F}_T{}^t \circ \mathbf{t}_s \right] \mathbf{t}_s{}^T \mathbf{S}_s \right. \\
& - \frac{p_s \varepsilon_{Tm} g_{Tmn}^e}{\left(F_T{}^t\right)^2 \left\| \mathbf{x}_{mn} - \mathbf{x}_{mpf} \right\|} \mathbf{F}_T{}^t \left(\mathbf{1} - \mathbf{t}_m \otimes \mathbf{t}_m\right) \mathbf{S}_m - \frac{p_s \varepsilon_{Tm}}{\left(F_T{}^t\right)^2} \left(\mathbf{F}_T{}^t \circ \mathbf{t}_m\right) \mathbf{t}_m{}^T \mathbf{S}_m \\
& \left. - \frac{p_s \varepsilon_{Ts} g_{Tsn}^e}{\left(F_T{}^t\right)^2 \left\| \mathbf{x}_{sn} - \mathbf{x}_{spf} \right\|} \mathbf{F}_T{}^t \left(\mathbf{1} - \mathbf{t}_s \otimes \mathbf{t}_s\right) \mathbf{S}_s \right\} \begin{bmatrix} \Delta \mathbf{u}_M \\ \Delta \mathbf{u}_S \end{bmatrix}.
\end{aligned} \tag{3.53}
$$

Defining the matrices in brackets in (3.53) as $\mathbf{P}_m$ and $\mathbf{P}_s$ one gets

$$
\begin{aligned}
\Delta p_m &= \mathbf{P}_m \begin{bmatrix} \Delta \mathbf{u}_M \\ \Delta \mathbf{u}_S \end{bmatrix}, \\
\Delta p_s &= \mathbf{P}_s \begin{bmatrix} \Delta \mathbf{u}_M \\ \Delta \mathbf{u}_S \end{bmatrix}.
\end{aligned} \tag{3.54}
$$

Now all the required variables are discretised. To simplify further the notation the following matrices are also introduced

$$
\begin{aligned}
\mathbf{Z}_m &= \mathbf{F}_m{}^T \mathbf{t}_m{}^T \left[\mathbf{H}_{mn} \quad \mathbf{0}\right], \\
\mathbf{Z}_s &= \mathbf{F}_s{}^T \mathbf{t}_s{}^T \left[\mathbf{0} \quad \mathbf{H}_{sn}\right],
\end{aligned} \tag{3.55}
$$

$$
\begin{aligned}
\mathbf{W}_m &= \begin{bmatrix} \sum_{j=1}^{3} t_{mj} \left(\mathbf{G}_{djmn} - \mathbf{G}_{djmp}\right) & \mathbf{0} \\ \mathbf{0} & \mathbf{0} \end{bmatrix}, \\
\mathbf{W}_s &= \begin{bmatrix} \mathbf{0} & \mathbf{0} \\ \mathbf{0} & \sum_{j=1}^{3} t_{sj} \left(\mathbf{G}_{djsn} - \mathbf{G}_{djsp}\right) \end{bmatrix},
\end{aligned} \tag{3.56}
$$

where the components of the tangent vectors (3.30) or (3.31) are used.

3.5 Residual Vector and Tangent Stiffness Matrix for Friction

The discretised kinematic variables presented in the previous section can be substituted to the appropriate expressions of the weak form and its linearisation presented in Section 3.3. In this way the residual vector and the tangent stiffness matrix related to friction for the beam-to-beam contact finite element can be derived. After adding them to the corresponding vectors and matrices for the normal contact presented in Section 2.6 the total vectors and matrices for frictional contact can be found.

In the penalty method for the stick state the friction contribution to the weak form (3.34) after the discretisation takes the form

$$\begin{aligned} \delta\Pi_{Tm}^{e\varepsilon} &= \left(\delta\mathbf{u}_M{}^T, \delta\mathbf{u}_S{}^T\right)\left[\mathbf{R}_{Tm}^{e\varepsilon}\right]_{24\times1}, \\ \delta\Pi_{Ts}^{e\varepsilon} &= \left(\delta\mathbf{u}_M{}^T, \delta\mathbf{u}_S{}^T\right)\left[\mathbf{R}_{Ts}^{e\varepsilon}\right]_{24\times1}, \end{aligned} \tag{3.57}$$

where the residual vectors can be written down as

$$\begin{aligned} \mathbf{R}_{Tm}^{e\varepsilon} &= \varepsilon_{Tm}\, g_{Tmn}^{e}\, \mathbf{R}_2, \\ \mathbf{R}_{Ts}^{e\varepsilon} &= \varepsilon_{Ts}\, g_{Tsn}^{e}\, \mathbf{R}_3 \end{aligned} \tag{3.58}$$

and the auxiliary vectors are

$$\begin{aligned} \mathbf{R}_2 &= \mathbf{S}_m{}^T\, \mathbf{t}_m, \\ \mathbf{R}_3 &= \mathbf{S}_s{}^T\, \mathbf{t}_s. \end{aligned} \tag{3.59}$$

For the slip state from the formulae (3.36) and (3.38) one gets

$$\begin{aligned} \delta\Pi_{Tm}^{p\varepsilon} &= \left(\delta\mathbf{u}_M{}^T, \delta\mathbf{u}_S{}^T\right)\left[\mathbf{R}_{Tm}^{p\varepsilon}\right]_{24\times1}, \\ \delta\Pi_{Ts}^{p\varepsilon} &= \left(\delta\mathbf{u}_M{}^T, \delta\mathbf{u}_S{}^T\right)\left[\mathbf{R}_{Ts}^{p\varepsilon}\right]_{24\times1} \end{aligned} \tag{3.60}$$

and the corresponding residual vectors for the case, when the friction is treated with two independent friction forces acting along two contacting beams, are

$$\begin{aligned} \mathbf{R}_{Tm}^{p\varepsilon} &= \mu\varepsilon_N g_N\, \mathbf{R}_2, \\ \mathbf{R}_{Ts}^{p\varepsilon} &= \mu\varepsilon_N g_N\, \mathbf{R}_3 \end{aligned} \tag{3.61}$$

while for the approach with one resultant friction force –

$$\begin{aligned} \mathbf{R}_{Tm}^{p\varepsilon} &= \mu\varepsilon_N g_N p_m s_m\, \mathbf{R}_2, \\ \mathbf{R}_{Ts}^{p\varepsilon} &= \mu\varepsilon_N g_N p_s s_s\, \mathbf{R}_3. \end{aligned} \tag{3.62}$$

In the stick state discretisation of the linearisation of the weak form (3.35) leads to the following relations

$$\Delta\delta\Pi_{Tm}^{e\varepsilon} = \left(\delta\mathbf{u}_M{}^T, \delta\mathbf{u}_S{}^T\right)\left[\mathbf{K}_{Tm}^{e\varepsilon}\right]_{24\times 24}\begin{bmatrix}\Delta\mathbf{u}_M\\ \Delta\mathbf{u}_S\end{bmatrix},$$
$$\Delta\delta\Pi_{Ts}^{e\varepsilon} = \left(\delta\mathbf{u}_M{}^T, \delta\mathbf{u}_S{}^T\right)\left[\mathbf{K}_{Ts}^{e\varepsilon}\right]_{24\times 24}\begin{bmatrix}\Delta\mathbf{u}_M\\ \Delta\mathbf{u}_S\end{bmatrix} \tag{3.63}$$

and to the corresponding tangent stiffness matrices

$$\mathbf{K}_{Tm}^{e\varepsilon} = \varepsilon_{Tm}\,\mathbf{R}_2 \otimes \mathbf{R}_2 + \varepsilon_{Tm}\, g_{Tmn}^{e}\,\mathbf{K}_3\,,$$
$$\mathbf{K}_{Ts}^{e\varepsilon} = \varepsilon_{Ts}\,\mathbf{R}_3 \otimes \mathbf{R}_3 + \varepsilon_{Ts}\, g_{Tsn}^{e}\,\mathbf{K}_4\,, \tag{3.64}$$

where the auxiliary matrices have the form

$$\mathbf{K}_3 = \frac{1}{d_{gTmn}}\mathbf{S}_m{}^T\left(\mathbf{1}-\mathbf{t}_m \otimes \mathbf{t}_m\right)\mathbf{S}_m + \mathbf{t}_m{}^T\,\mathbf{x}_{mn,mm}\,\mathbf{F}_m{}^T\,\mathbf{F}_m + \mathbf{Z}_m + \mathbf{Z}_m{}^T$$
$$-\mathbf{t}_m{}^T\,\mathbf{x}_{mn,m}\left(a_{mm}\mathbf{R}_m + a_{ms}\mathbf{R}_s\right) + \mathbf{W}_m\,,$$
$$\mathbf{K}_4 = \frac{1}{d_{gTsn}}\mathbf{S}_s{}^T\left(\mathbf{1}-\mathbf{t}_s \otimes \mathbf{t}_s\right)\mathbf{S}_s + \mathbf{t}_s{}^T\,\mathbf{x}_{sn,ss}\,\mathbf{F}_s{}^T\,\mathbf{F}_s + \mathbf{Z}_s + \mathbf{Z}_s{}^T$$
$$-\mathbf{t}_s{}^T\,\mathbf{x}_{sn,s}\left(a_{sm}\mathbf{R}_m + a_{ss}\mathbf{R}_s\right) + \mathbf{W}_s\,. \tag{3.65}$$

In Eq. (3.65) the decomposition of the inverse of the matrix **A** into its elements as in (A2.4) is used.

For the slip state from the formulae (3.37) and (3.39) one gets

$$\Delta\delta\Pi_{Tm}^{p\varepsilon} = \left(\delta\mathbf{u}_M{}^T, \delta\mathbf{u}_S{}^T\right)\left[\mathbf{K}_{Tm}^{p\varepsilon}\right]_{24\times 24}\begin{bmatrix}\Delta\mathbf{u}_M\\ \Delta\mathbf{u}_S\end{bmatrix},$$
$$\Delta\delta\Pi_{Ts}^{p\varepsilon} = \left(\delta\mathbf{u}_M{}^T, \delta\mathbf{u}_S{}^T\right)\left[\mathbf{K}_{Ts}^{p\varepsilon}\right]_{24\times 24}\begin{bmatrix}\Delta\mathbf{u}_M\\ \Delta\mathbf{u}_S\end{bmatrix} \tag{3.66}$$

and the appropriate tangent stiffness matrices have the form

$$\mathbf{K}_{Tm}^{p\varepsilon} = \mu\varepsilon_N \mathbf{R}_2 \otimes \mathbf{R}_1 + \mu\varepsilon_N g_N \mathbf{K}_3\,,$$
$$\mathbf{K}_{Ts}^{p\varepsilon} = \mu\varepsilon_N\,\mathbf{R}_3 \otimes \mathbf{R}_1 + \mu\varepsilon_N g_N \mathbf{K}_4 \tag{3.67}$$

for the separate treatment of friction forces or

$$\mathbf{K}_{Tm}^{p\varepsilon} = \mu\varepsilon_N p_m s_m\,\mathbf{R}_2 \otimes \mathbf{R}_1 + \mu\varepsilon_N g_N p_m s_m \mathbf{K}_3 + \mu\varepsilon_N g_N s_m\,\mathbf{R}_2 \otimes \mathbf{P}_m\,,$$
$$\mathbf{K}_{Ts}^{p\varepsilon} = \mu\varepsilon_N p_s s_s\,\mathbf{R}_3 \otimes \mathbf{R}_1 + \mu\varepsilon_N g_N p_s s_s \mathbf{K}_4 + \mu\varepsilon_N g_N s_s \mathbf{R}_3 \otimes \mathbf{P}_s \tag{3.68}$$

for the friction forces considered as the components of one resultant friction force.

The vector $\mathbf{R}_1$ present in (3.67) and (3.68) was defined in Section 2.6 by (2.45).

In the Lagrange multipliers metod, after introduction of friction, the vector of unknowns has three more elements – one Lagrange multiplier for the normal contact – λ_N, and two multipliers for friction – λ_{Tm} and λ_{Ts}. Hence, in the stick state the influences of friction in the weak form (3.40) after discretisation can be expressed as

$$\begin{aligned}\delta\Pi_{Tm}^{e\lambda} &= \left(\delta\mathbf{u}_M{}^T, \delta\mathbf{u}_S{}^T, \delta\lambda_N, \delta\lambda_{Tm}, \delta\lambda_{Ts}\right)\left[\mathbf{R}_{Tm}^{e\lambda}\right]_{27\times1},\\ \delta\Pi_{Ts}^{e\lambda} &= \left(\delta\mathbf{u}_M{}^T, \delta\mathbf{u}_S{}^T, \delta\lambda_N, \delta\lambda_{Tm}, \delta\lambda_{Ts}\right)\left[\mathbf{R}_{Ts}^{e\lambda}\right]_{27\times1},\end{aligned} \tag{3.69}$$

where the residual vectors have the form

$$\mathbf{R}_{Tm}^{e\lambda} = \begin{bmatrix}\lambda_{Tm}\mathbf{R}_2\\ 0\\ g_{Tmn}\\ 0\end{bmatrix},\qquad \mathbf{R}_{Ts}^{e\lambda} = \begin{bmatrix}\lambda_{Ts}\mathbf{R}_3\\ 0\\ 0\\ g_{Tsn}\end{bmatrix} \tag{3.70}$$

with the auxiliary vectors defined in (3.59).

In the slip state after discretisation of the expression (3.42) or (3.44) one gets

$$\begin{aligned}\delta\Pi_{Tm}^{p\lambda} &= \left(\delta\mathbf{u}_M{}^T, \delta\mathbf{u}_S{}^T, \delta\lambda_N, \delta\lambda_{Tm}, \delta\lambda_{Ts}\right)\left[\mathbf{R}_{Tm}^{p\lambda}\right]_{27\times1},\\ \delta\Pi_{Ts}^{p\lambda} &= \left(\delta\mathbf{u}_M{}^T, \delta\mathbf{u}_S{}^T, \delta\lambda_N, \delta\lambda_{Tm}, \delta\lambda_{Ts}\right)\left[\mathbf{R}_{Ts}^{p\lambda}\right]_{27\times1}\end{aligned} \tag{3.71}$$

and the corresponding residual vectors have the form

$$\mathbf{R}_{Tm}^{p\lambda} = \begin{bmatrix}\mu\lambda_N\mathbf{R}_2\\ 0\\ |\lambda_{Tm}| - \mu\lambda_N\\ 0\end{bmatrix},\qquad \mathbf{R}_{Ts}^{p\lambda} = \begin{bmatrix}\mu\lambda_N\mathbf{R}_3\\ 0\\ 0\\ |\lambda_{Ts}| - \mu\lambda_N\end{bmatrix} \tag{3.72}$$

for separate treatment of friction forces or

$$\mathbf{R}_{Tm}^{p\lambda} = \begin{bmatrix} \mu\lambda_N s_m p_m \mathbf{R}_2 \\ 0 \\ |\lambda_{Tm}| - \mu p_m \lambda_N \\ 0 \end{bmatrix},$$
$$\mathbf{R}_{Ts}^{p\lambda} = \begin{bmatrix} \mu\lambda_N s_s p_s \mathbf{R}_3 \\ 0 \\ 0 \\ |\lambda_{Ts}| - \mu p_s \lambda_N \end{bmatrix}, \tag{3.73}$$

when the forces along two contacting beams are considered as the components of one resultant friction force.

In the stick state discretisation of the linearization of the weak form (3.41) leads to

$$\Delta\delta\Pi_{Tm}^{e\lambda} = \left(\delta\mathbf{u}_M{}^T, \delta\mathbf{u}_S{}^T, \delta\lambda_N, \delta\lambda_{Tm}, \delta\lambda_{Ts}\right)\left[\mathbf{K}_{Tm}^{e\lambda}\right]_{27\times 27} \begin{bmatrix} \Delta\mathbf{u}_M \\ \Delta\mathbf{u}_S \\ \Delta\lambda_N \\ \Delta\lambda_{Tm} \\ \Delta\lambda_{Ts} \end{bmatrix} \tag{3.74$_1$}$$

$$\Delta\delta\Pi_{Ts}^{e\lambda} = \left(\delta\mathbf{u}_M{}^T, \delta\mathbf{u}_S{}^T, \delta\lambda_N, \delta\lambda_{Tm}, \delta\lambda_{Ts}\right)\left[\mathbf{K}_{Ts}^{e\lambda}\right]_{27\times 27} \begin{bmatrix} \Delta\mathbf{u}_M \\ \Delta\mathbf{u}_S \\ \Delta\lambda_N \\ \Delta\lambda_{Tm} \\ \Delta\lambda_{Ts} \end{bmatrix} \tag{3.74$_2$}$$

and the appropriate tangent stiffness matrices are

$$\mathbf{K}_{Tm}^{e\lambda} = \begin{bmatrix} \lambda_{Tm}\mathbf{K}_3 & \mathbf{0} & \mathbf{R}_2 & \mathbf{0} \\ \mathbf{0}^T & 0 & 0 & 0 \\ \mathbf{R}_2{}^T & 0 & 0 & 0 \\ \mathbf{0}^T & 0 & 0 & 0 \end{bmatrix},$$
$$\mathbf{K}_{Ts}^{e\lambda} = \begin{bmatrix} \lambda_{Ts}\mathbf{K}_4 & \mathbf{0} & \mathbf{0} & \mathbf{R}_3 \\ \mathbf{0}^T & 0 & 0 & 0 \\ \mathbf{0}^T & 0 & 0 & 0 \\ \mathbf{R}_3{}^T & 0 & 0 & 0 \end{bmatrix}. \tag{3.75}$$

In the slip state after discretisation of the formulae (3.43) or (3.45) one gets

$$\Delta\delta\Pi_{Tm}^{p\lambda} = \left(\delta\mathbf{u}_M{}^T, \delta\mathbf{u}_S{}^T, \delta\lambda_N, \delta\lambda_{Tm}, \delta\lambda_{Ts}\right)\left[\mathbf{K}_{Tm}^{p\lambda}\right]_{27\times27} \begin{bmatrix} \Delta\mathbf{u}_M \\ \Delta\mathbf{u}_S \\ \Delta\lambda_N \\ \Delta\lambda_{Tm} \\ \Delta\lambda_{Ts} \end{bmatrix},$$
$$\Delta\delta\Pi_{Ts}^{p\lambda} = \left(\delta\mathbf{u}_M{}^T, \delta\mathbf{u}_S{}^T, \delta\lambda_N, \delta\lambda_{Tm}, \delta\lambda_{Ts}\right)\left[\mathbf{K}_{Ts}^{p\lambda}\right]_{27\times27} \begin{bmatrix} \Delta\mathbf{u}_M \\ \Delta\mathbf{u}_S \\ \Delta\lambda_N \\ \Delta\lambda_{Tm} \\ \Delta\lambda_{Ts} \end{bmatrix} \tag{3.76}$$

and the tangent stiffness matrices have the form

$$\mathbf{K}_{Tm}^{p\lambda} = \begin{bmatrix} \mu\lambda_N\mathbf{K}_3 & \mu\mathbf{R}_2 & \mathbf{0} & \mathbf{0} \\ \mathbf{0}^T & 0 & 0 & 0 \\ \mathbf{0}^T & -\mu & s_m & 0 \\ \mathbf{0}^T & 0 & 0 & 0 \end{bmatrix},$$
$$\mathbf{K}_{Ts}^{p\lambda} = \begin{bmatrix} \mu\lambda_N\mathbf{K}_4 & \mu\mathbf{R}_3 & \mathbf{0} & \mathbf{0} \\ \mathbf{0}^T & 0 & 0 & 0 \\ \mathbf{0}^T & 0 & 0 & 0 \\ \mathbf{0}^T & -\mu & 0 & s_s \end{bmatrix}, \tag{3.77}$$

when the friction forces are treated separately or

$$\mathbf{K}_{Tm}^{p\lambda} = \begin{bmatrix} \mu\lambda_N s_m\left(p_m\mathbf{K}_3 + \mathbf{R}_2\otimes\mathbf{P}_m\right) & \mu p_m s_m\mathbf{R}_2 & \mathbf{0} & \mathbf{0} \\ \mathbf{0}^T & 0 & 0 & 0 \\ \mathbf{0}^T & -\mu p_m & 1 & 0 \\ \mathbf{0}^T & 0 & 0 & 0 \end{bmatrix},$$
$$\mathbf{K}_{Ts}^{p\lambda} = \begin{bmatrix} \mu\lambda_N s_s\left(p_s\mathbf{K}_4 + \mathbf{R}_3\otimes\mathbf{P}_m\right) & \mu p_s s_s\mathbf{R}_3 & \mathbf{0} & \mathbf{0} \\ \mathbf{0} & 0 & 0 & 0 \\ \mathbf{0}^T & 0 & 0 & 0 \\ \mathbf{0}^T & -\mu p_s & 0 & 1 \end{bmatrix}, \tag{3.78}$$

when they are considered as the components of one resultant friction force.

In the residual vectors and the tangent stiffness matrices for the Lagrange multipliers method in the slip state, besides the terms resulting directly from the weak form and its linearisation, additional expressions, which ensure fulfillment of conditions (3.16) and (3.20) are present. In this way limiting of the maximal value of the friction force in the case of sliding, resulting from Coulomb friction law, is automatically fulfilled.

It should also be pointed out, that when the friction forces in the directions of both contacting beams are treated independently, two different friction states, stick or slip, are allowed in one contact point. In such a case the entire residual vector and the tangent stiffness matrix should be composed as a sum of appropriate components presented above. On the other hand, when the friction forces along both beams are considered as components of one resultant friction force, then one contact point is characterized by only one friction state – stick or slip. In this situation separation of the friction contributions into parts for beams *m* and *s* is not required. Hence, the residual vector for the penalty method is a sum of $(3.58)_1$ and $(3.58)_2$ in the stick state or a sum of $(3.62)_1$ and $(3.62)_2$ in the slip state. In the Lagrange multipliers method for the stick state one can add $(3.70)_1$ and $(3.70)_2$, while for the slip state – $(3.73)_1$ and $(3.73)_2$. Similarly, one can add appropriate tangent stiffness matrices: in the penalty method for the stick state – $(3.64)_1$ and $(3.64)_2$ or for the slip state – $(3.68)_1$ and $(3.68)_2$, while for the Lagrange multipliers method for the stick state – $(3.75)_1$ and $(3.75)_2$ or for the slip state – $(3.78)_1$ and $(3.78)_2$.

3.6 Numerical Examples

3.6.1 Introduction

In this section five examples of beam-to-beam frictional contact analysis are presented. In all the examples the friction forces along the contacting beams are treated independently. The second approach with one resultant friction force is used in examples presented in Chapter 4.

The main purpose of the presented examples is a comparison of the penalty method and the Lagrange multipliers method. Accuracy and computer time of calculations are considered. Results in the first example are also compared with numerical results for the full 3D analysis of contacting solids.

The computer time depends, among others, on the code length. The *exe*-type file obtained using the Compaq Visual Fortran compiler has 749 kB for the penalty method and 777 kB for the Lagrange multipliers method (Litewka 2005). The difference in length is due to the necessary additional checks and precautions resulting from the exact fulfillment of constraints. These problems were mentioned in previous sections. Besides, the Lagrange multipliers method involves more unknowns and the expected computer time is longer. Additionally, in the presented examples the influence of the penalty parameter and the friction coefficient on the results is presented.

In the examples loading is applied in the form of displacements imposed simultaneously in a given number of equal increments. The loading and the deformation process is measured in fictitious time and a given moment in this time is determined by the parameter T, which goes from 0 to 1.

Each of the examples is solved using the Newton-Raphson method. The convergence criterion uses the relative energy, which is calculated as a scalar product of the residual vector and the displacement increment vector related to such a product for the first iteration in a given increment. The convergence is assumed if such a relative energy is smaller than 10^{-8}.

In all the examples each of the beams is discretised with 10 identical beam finite elements, according to the conclusion from Section 2.7.2.

The data used in the examples are devoid of any physical units. It is understood, however, that any system of physically consistent units might be attached.

3.6.2 Example 1

In this example contact between two beams shown in Fig. 3.2 is analysed. Beam **1** is a cantilever and beam **2** has the boundary conditions resulting from its symmetry in its centre point. Besides its ends have the restrained rotation about the longitudinal axis of the beam – Y. At these ends imposed displacements, $\Delta_1 = 40$ and

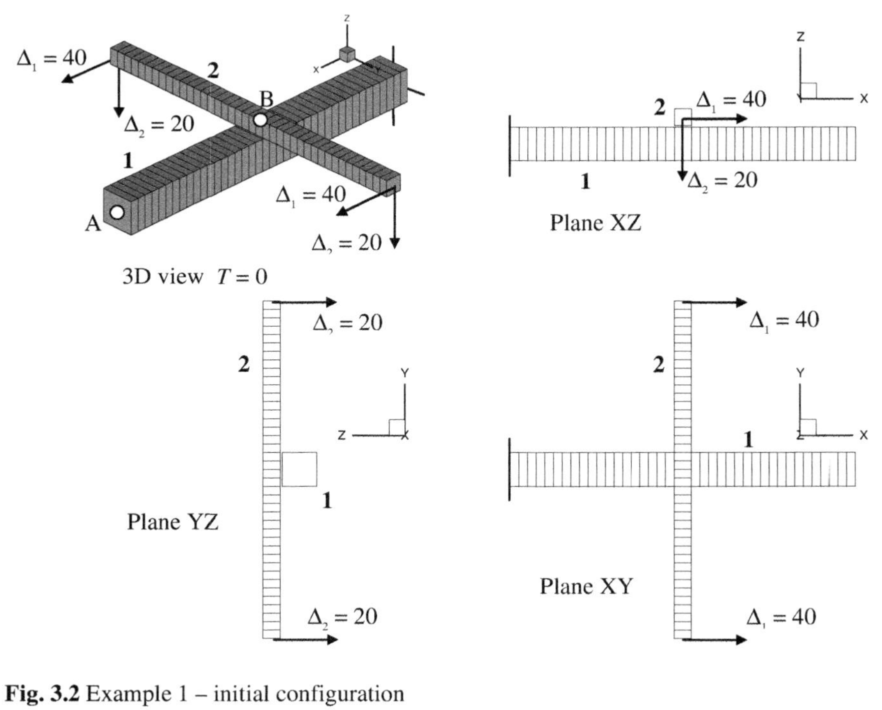

Fig. 3.2 Example 1 – initial configuration

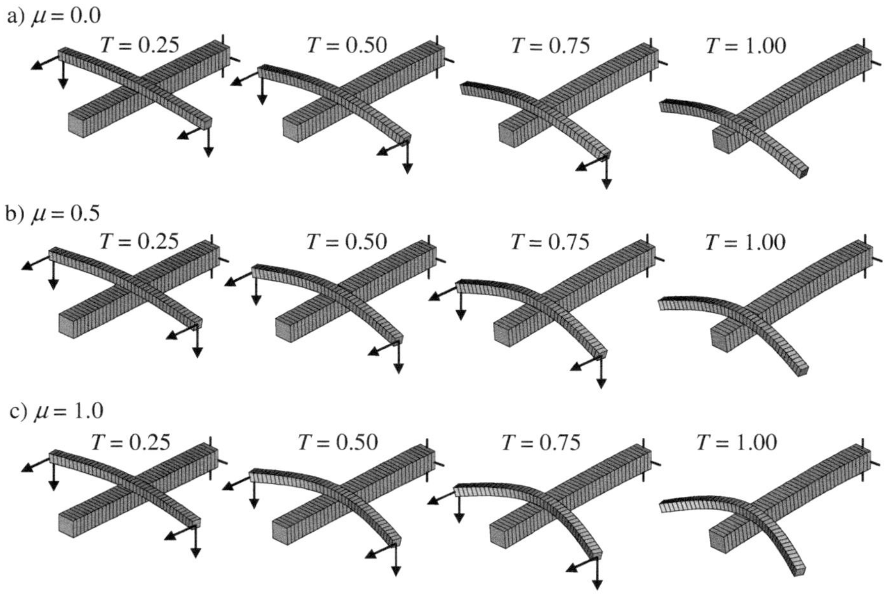

Fig. 3.3 Example 1 – deformed configurations of beams for: a) μ = 0.0, b) μ = 0.5, c) μ = 1.0

Table 3.1 Example 1 – friction states

μ	Friction state on beam	*T*			
		0.25	0.50	0.75	1.00
0.0	**1**	slip	slip	slip	slip
	2	slip	slip	slip	slip
0.5	**1**	slip	slip	slip	slip
	2	stick	stick	stick	stick
1.0	**1**	stick	slip	slip	slip
	2	stick	stick	stick	stick

$\Delta_2 = 20$, are applied in 50 increments. The following data are taken to the calculations: beam **1** – $E = 30000$, $\nu = 0.17$, cross-section dimensions $b_s = b_t = 10$, length 100; beam **2** – $E = 30000$, $\nu = 0.17$, cross-section dimensions $b_s = b_t = 5$, length 100; initial gap between the beams 0.5 and three values of friction coefficient $\mu = 0.0$; 0.5 or 1.0.

Deformed configurations of the beams in four selected moments of the deformation process for three analysed values of the friction coefficient are presented in Fig. 3.3. These are results obtained using the penalty method with the penalty parameters $\varepsilon_N = 25000$ and $\varepsilon_{Tm} = \varepsilon_{Ts} = 250$. Besides, in Table 3.1 the states of friction

Table 3.2 Example 1 – displacements at points A and B

Calculation method		Point A		Point B	
Penalty		δ_X	δ_Z	δ_X	δ_Z
ε_N	$\varepsilon_{Tm} = \varepsilon_{Ts}$				
10000	400	0.537	9.79	8.02	35.6
10000	800	0.538	9.79	8.02	35.5
10000	1500	0.539	9.80	8.02	35.5
20000	400	0.539	9.81	8.01	35.6
20000	800	0.539	9.82	8.01	35.5
20000	1500	0.540	9.82	8.01	35.5
30000	400	0.539	9.80	8.00	35.6
30000	800	0.540	9.81	8.00	35.5
30000	1500	0.541	9.82	8.00	35.4
Lagrange multipliers		0.571	9.89	8.08	36.2
3D analysis, Abaqus		0.562	9.87	8.27	37.2

corresponding to the deformed configurations in Fig. 3.3 are given. These results exhibit a clear influence of the friction coefficient on the friction state, and therefore also on the beams deformations. The case of $\mu = 0.0$ is characterized throughout by the slip state and an increase of the friction coefficient delays appearance of this state. However, even when the sliding starts, the larger value of this coefficient prevents to a greater extent this relative displacement, what is manifested in a larger flexural deformation of beam **2**.

Table 3.2 contains a comparison of displacements in the directions of the axes X and Z for the points A and B depicted in Fig. 3.2. These values were calculated using the penalty method for various penalty parameters ε_N and $\varepsilon_{Tm} = \varepsilon_{Ts}$, the Lagrange multipliers method and the full 3D analysis, where the beams were modelled as solids. To this end the program Abaqus was used. Beam **1** was considered as a master body and discretised with 2x2x20 identical elements of C3D8 type. For beam **2**, which was a slave body, a finer mesh of 2x2x40 identical finite elements of the same type was applied.

The presented results do not show much differences. However, according to the expectations, the Lagrange multipliers method gives the displacements, which are closer to the 3D solution than the penalty method. This can be attributed to the fact, that the former method ensures the exact fulfilment of constraints. But the analysis of costs of this better accuracy is also important. The computer time on PC with a 900 MHz processor for the Lagrange multipliers method was 87 seconds, while for the penalty method – only 37 seconds.

It can also be observed that the influence of the penalty parameters values in the considered range on the results is negligible.

It is difficult to uniquely conclude, which of the methods is better. The presented example confirmed very well known advantages and disadvantages of them. The choice is rather dependent on the preferences of the user and on the type of analysis. The necessity to select properly the values of the penalty parameters requires certain experience and it can well happen, that the solution process must be repeated several times, until the satisfactory compromise between the acceptable violation of the constraint conditions and the ill-posedness of the matrices is reached. On the other hand, the better accuracy of the Lagrange multipliers method is bought with a more complicated computer code and a larger number of unknowns. This last feature can be especially costly, when the contact with a large number of contact points is analysed.

3.6.3 Example 2

In this example contact between two cantilever beams shown in Fig. 3.4 is analysed. The ends of beam **2** are subjected to displacements $\Delta_1 = 20$ and $\Delta_2 = 20$ applied in 40 increments. The following data are taken to the calculations: beam **1** – $E = 20000$, $\nu = 0.3$, cross-section dimensions $b_s = b_t = 10$, length 100; beam **2** – $E = 30000$, $\nu = 0.17$, cross-section dimensions $b_s = b_t = 5$, length 100; initial gap between the beams – 0.15 and the friction coefficient $\mu = 1.0$.

Deformed configurations of the beams in four selected moments of the deformation process are presented in Fig. 3.5. They are obtained using the penalty method with $\varepsilon_N = 12500$ and $\varepsilon_{Tm} = \varepsilon_{Ts} = \varepsilon_T = 1000$. In this case the friction in contact on both beams is in the stick state.

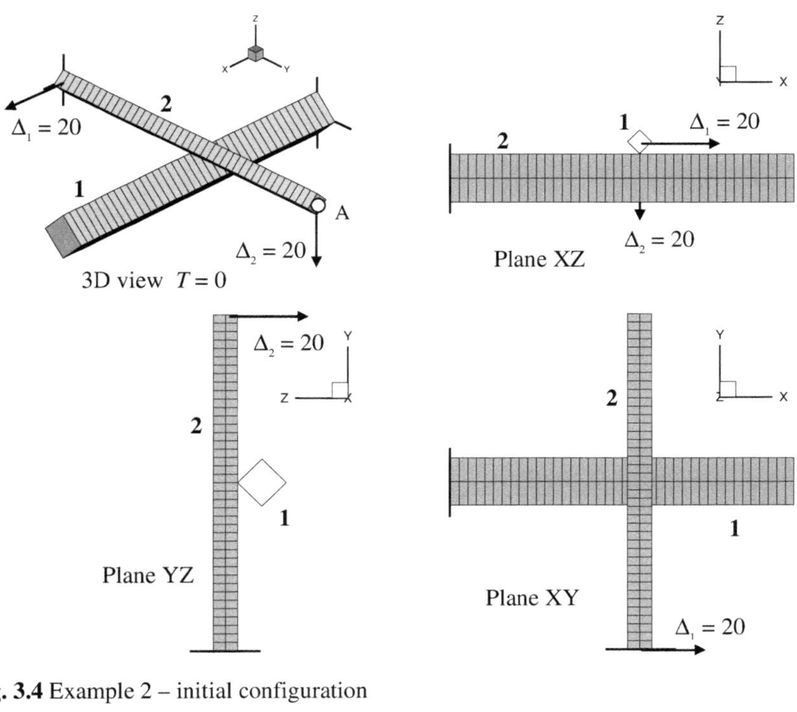

Fig. 3.4 Example 2 – initial configuration

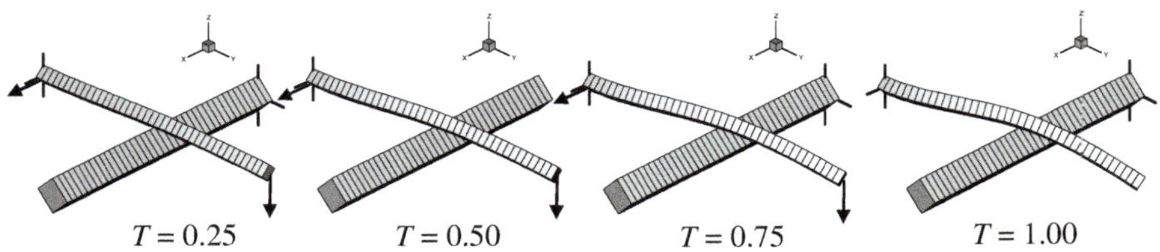

Fig. 3.5 Example 2 – deformed configurations of beams

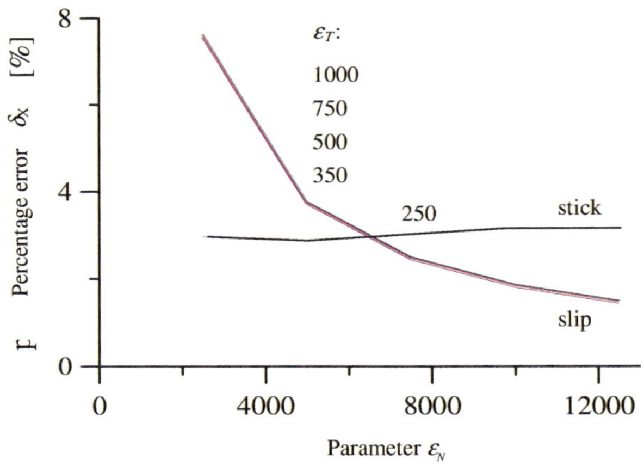

Fig. 3.6 Example 2 – percentage error for displacement at point A

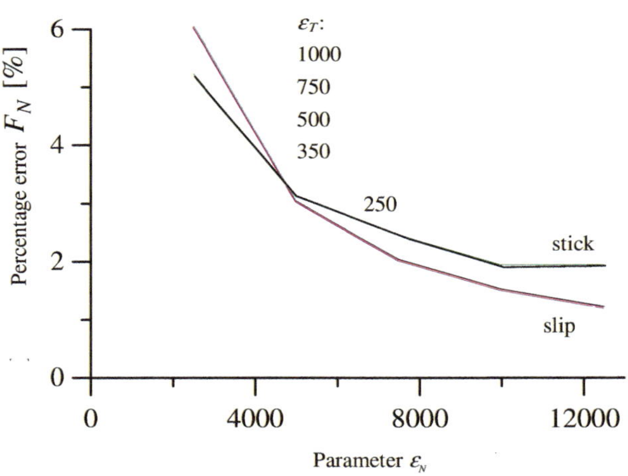

Fig. 3.7 Example 2 – percentage error for normal force in contact

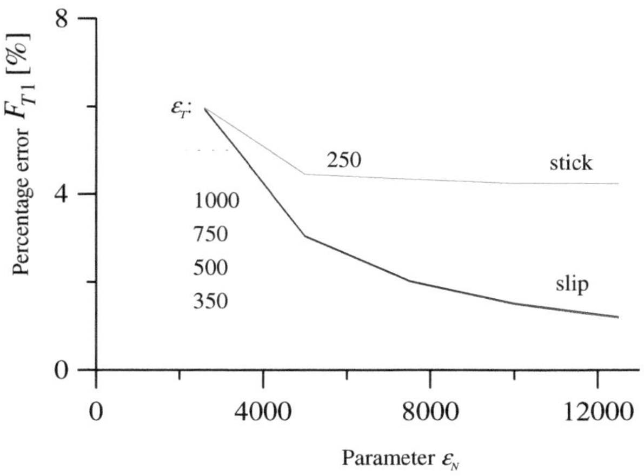

Fig. 3.8 Example 2 – percentage error for friction force in contact

Figures 3.6, 3.7 and 3.8 present results of influence of the penalty parameters values on the displacements along the axis X at the point A shown in Fig. 3.4 and the normal and friction forces at the contact point. The graphs show the percentage error of the respective results related to the values obtained by the Lagrange multipliers method, which are considered as reference.

It can be observed, that generally all the results concerning both the displacements and the forces are more sensitive to changes of the normal penalty parameter ε_N than the tangential one ε_T. It can be noted, though, that there exists a limiting value of the latter parameter, in the range between 250 and 350, for which the friction state in contact changes from stick to slip. Hence, care is required when adopting the proper value of this parameter, because otherwise the state of friction may be assessed incorrectly. If the value of the tangential penalty parameter is taken above this limit, then the penalty method allows to calculate of the characteristic contact variables with an error below 2%, what can be considered as a satisfactory accuracy.

The computer time in this example was 35 seconds for the penalty method and 50 seconds for the Lagrange multipliers method. The conclusions from these analyses coincide with that in Example 1 – the choice of the method is very subjective, it seems, that the advantages and disadvantages are balanced.

3.6.4 Example 3

In this example contact between two cantilever beams shown in Fig. 3.9 is analysed. The free end of beam **2** – point A, is subjected to a displacement $\Delta = 40$ applied in 40 increments. The following data are taken to the calculations: beam **1** – $E = 20000$, $\nu = 0.3$, cross-section dimensions $b_s = b_t = 10$, length 100; beam **2** – $E = 30000$, $\nu = 0.17$, cross-section dimensions $b_s = b_t = 5$, length 100; initial gap between the beams 1.393 and the friction coefficient $\mu = 1.0$.

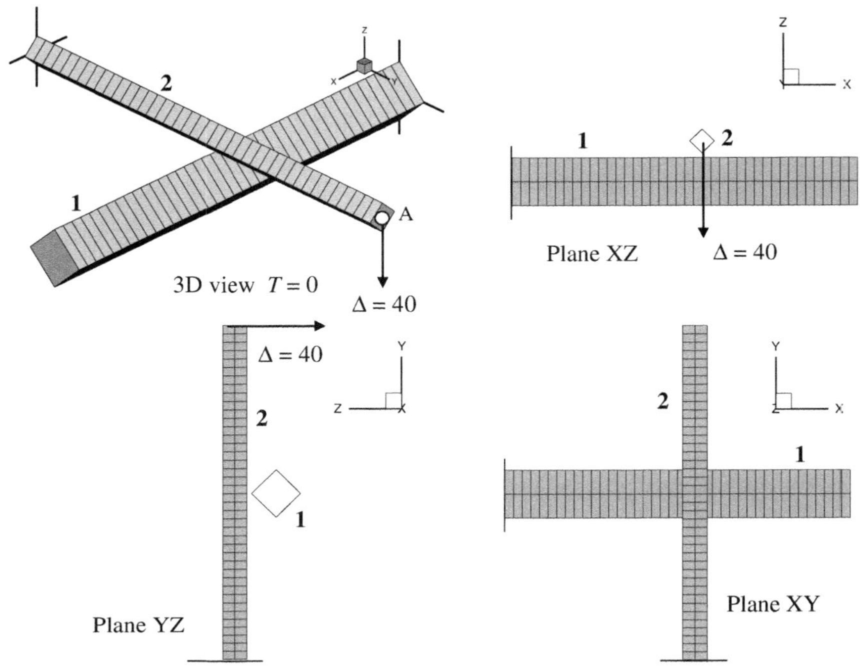

Fig. 3.9 Example 3 – initial configuration

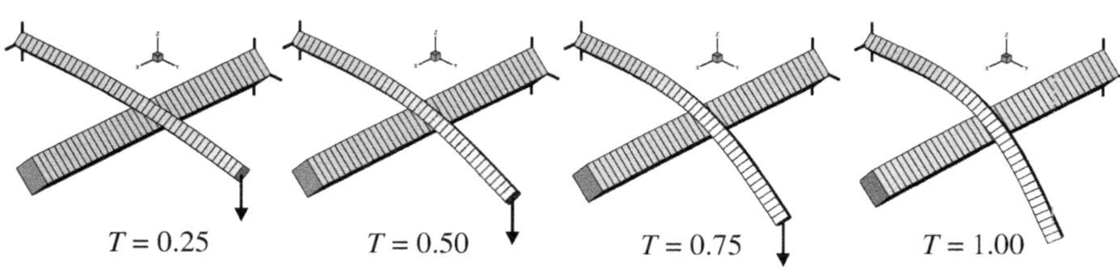

Fig. 3.10 Example 3 – deformed configurations of beams

Deformed configurations of the beams in four selected moments of the deformation process, which features the stick state throughout, are presented in Fig. 3.10. They are obtained using the penalty method.

Graphs in Figs. 3.11, 3.12 and 3.13 present an influence of the penalty parameters on the percentage error of the displacement at the point A in the direction of the axis X, the normal force and the friction force in contact. These values are related to the reference values obtained using the Lagrange multipliers method.

The results presented in Fig. 3.11 indicate, that in this example parameter ε_T has a significant influence on the displacement at point A. On the other hand, the influence of the parameter ε_N is negligible. However, it is not irrelevant at all – the

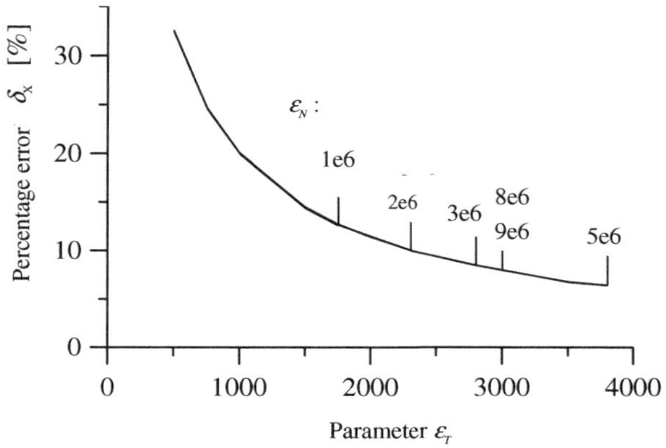

Fig. 3.11 Example 3 – percentage error for displacement at point A

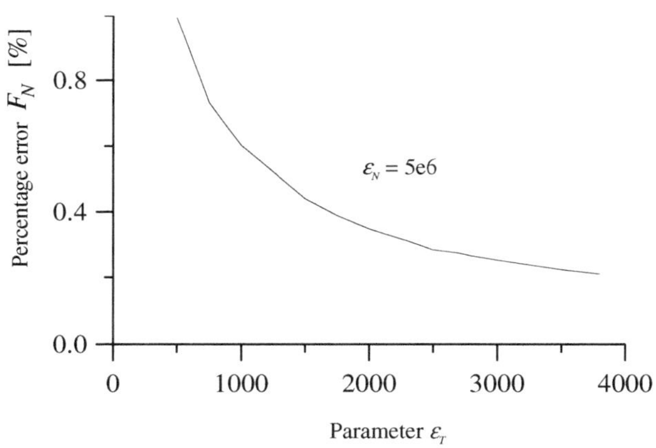

Fig. 3.12 Example 3 – percentage error for normal force in contact

vertical lines over the graph point the maximal values of ε_T, corresponding to given values of ε_N, for which the convergence of the iterative solution process still could be achieved. With ε_T exceeding those limits the ill-posedness of the matrices did not allow to reach the equilibrium state. It is also interesting to note that the optimal value of ε_N in this respect is not its maximal allowable $9 \cdot 10^6$ but rather $5 \cdot 10^6$, instead. This indicates a complexity of the contact phenomena and a complicated interplay between normal and friction forces. Their values presented in two subsequent figures are calculated for this optimal value of the parameter ε_N.

Observing the results in Fig. 3.12 it can be concluded that the normal force in the contact point can be found with a very good accuracy within 1% error even

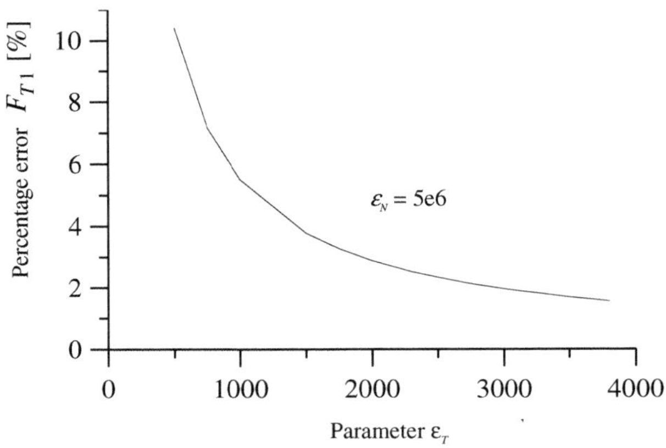

Fig. 3.13 Example 3 – percentage error for friction force in contact

with not optimally chosen values of the penalty parameters. However, their influence on the friction force is more significant. The graph in Fig. 3.13 shows, that the maximal possible penalty parameters allow to calculate the friction force with an error exceeding 2%.

The computer time for the Lagrange multipliers method in this case was 36 seconds and for the penalty method – 24 seconds.

It might be once again stated that the choice of the better method is not easy and the decision should depend on the type of example, user's experience and required accuracy of results.

3.6.5 Example 4

In this example contact between three cantilever beams shown in Fig. 3.14 is analysed. The free end of the beam **2** is subjected to displacement $\Delta = 20$ applied in 80 increments. This displacement starts a domino-like effect and leads finally to a mutual contact of all beams. The following data, identical for all beams, are taken to the calculations: $E = 30000$, $\nu = 0.3$, cross-section dimensions $b_s = b_t = 5$, length 100; initial gaps between beams **1** and **2** as well as **1** and **3** – 0.5; the friction coefficient $\mu = 0.0$ or 1.0. The example is solved using the penalty method with the following values of the penalty parameters: $\varepsilon_{N12} = 7000$, $\varepsilon_{N23} = 3000$, $\varepsilon_{Tm12} = \varepsilon_{Ts12} = \varepsilon_{Tm23} = \varepsilon_{Ts23} = 100$.

Deformed configurations of the beams for four selected moments of the deformation process and for two analyzed values of the friction coefficient are presented in Fig. 3.15. The influence of friction is clearly visible. The large deformation of the beams for the frictionless case results from unrestrained sliding between their edges.

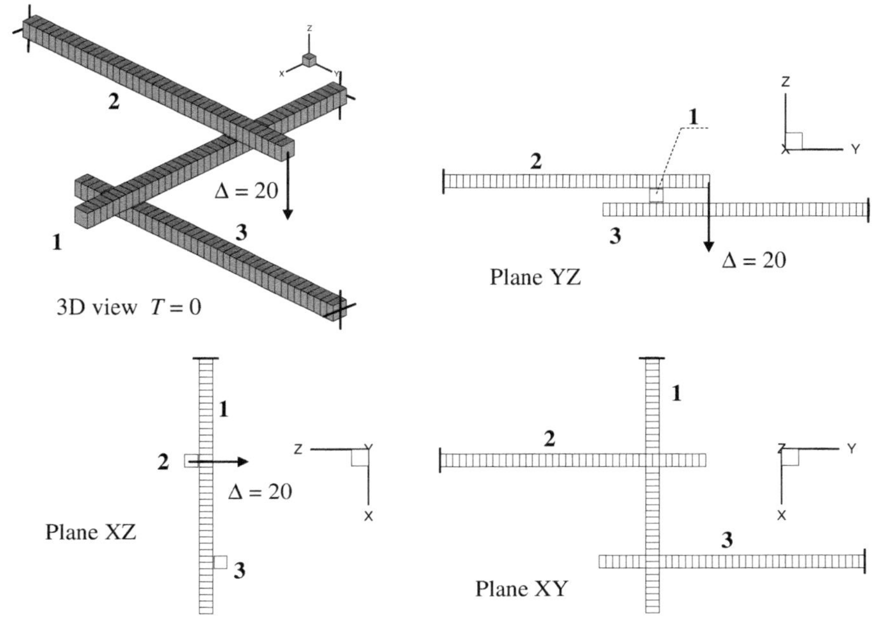

Fig. 3.14 Example 4 – initial configuration

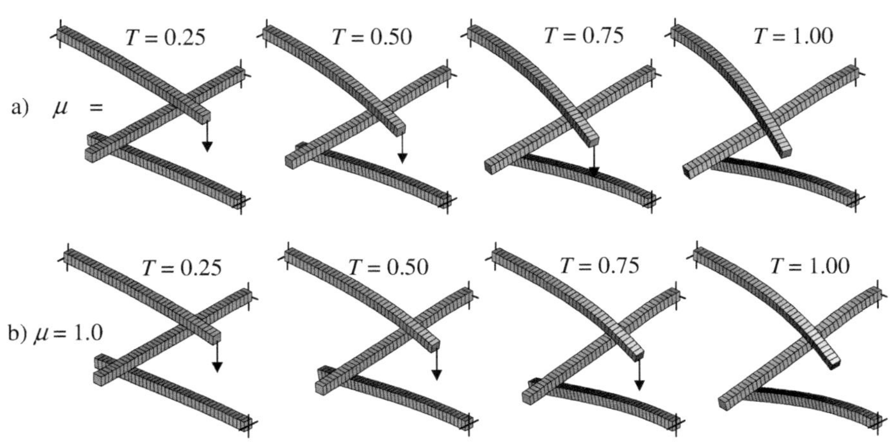

Fig. 3.15 Example 4 – deformed configurations of beams for: a) $\mu = 0.0$, b) $\mu = 1.0$

In Table 3.3 the friction states for the case with friction ($\mu = 1.0$) for the respective contact points at the moments of the deformation process corresponding to the graphical representation in Fig. 3.15b are presented. It can be seen, that during the

Table 3.3 Example 4 – friction state

State on beam	T			
	0.25	0.50	0.75	1.00
1 (with **2**)	stick	stick	stick	stick
2	stick	stick	stick	slip
1 (with **3**)	stick	stick	stick	stick
3	stick	stick	slip	slip

process the friction state in contact between beams **1** and **3** changes from stick to slip, while for the contact between beams **1** and **2** stick remains.

3.6.6 Example 5

In this example contact between two cantilever beams, shown in Fig. 3.16 is analysed. In the partially flexible support of beam **2** an elastic constraint of stiffness $k = 100$ is related to the angle of rotation about the longitudinal axis of the beam (X) and the displacement along axis Y is unconstrained. The free end of this beam is subjected to displacement $\Delta = 20$ applied in 50 increments.

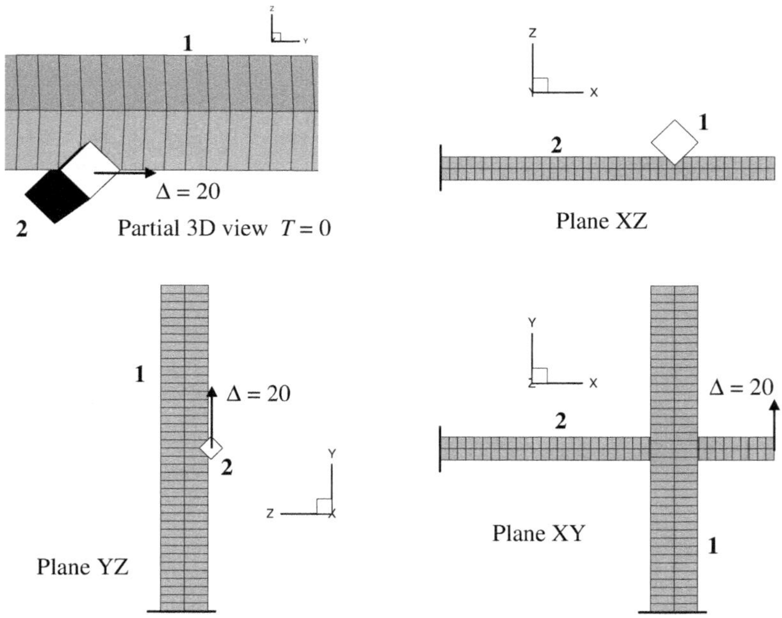

Fig. 3.16 Example 5 – initial configuration

The following data are taken to the calculations: beam **1** – $E = 30000$, $\nu = 0.17$, cross-section dimensions $b_s = b_t = 10$, length 100; beam **2** – $E = 10000$, $\nu = 0.17$, cross-section dimensions $b_s = b_t = 5$, length 100; initial penetration 2.6 and the friction coefficient $\mu = 0.0$ or 1.0. The example is solved using the penalty method with: $\varepsilon_N = 300$, $\varepsilon_{Tm} = \varepsilon_{Ts} = 20$.

The partial views of the deformed configurations of the beams in selected moments of the deformation process are presented in Fig. 3.17. In this example the influence of the friction coefficient on the results is clear as well. In the case without friction, after the initial penetration is eliminated, beam **2** undergoes sliding along beam **1**. On the other hand, in the frictional case, beam **2** initially rotates about the stationary contact point and only about $T = 0.2$, when the friction force assumes the limit value resulting from the Coulomb law, sliding starts. The friction states corresponding to the moments in the deformation process presented in Fig. 3.17 are summarized in Table 3.4.

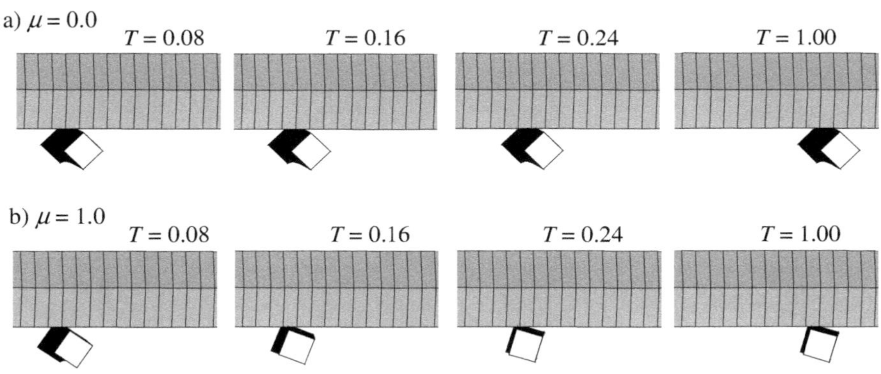

Fig. 3.17 Example 5 – partial views of deformed configurations of beams for: a) $\mu = 0.0$, b) $\mu = 1.0$

Table 3.4 Example 5 – friction states for $\mu = 1.0$

State on beam	T			
	0.08	0.16	0.24	1.00
1	stick	stick	stick	stick
2	stick	stick	slip	slip

In this example the choice of suitable values of the penalty parameters was additionally complicated by their interplay with the stiffness of the flexible support of beam **2**.

Chapter 4
Contact between Smoothed Beams

4.1 General Remarks on Smoothing of Contact Facets

A proper representation of surface geometry for bodies undergoing contact and sliding is a very important aspect of modeling in the finite element method. The fundamental issue here is to ensure continuity and smoothness. With a too simplified mesh of finite elements, definitions of normal and tangent vectors are not continuous between the adjacent elements. As was shown in Chapters 2 and 3, these vectors play the key role in the contact formulation. Hence, in the case when a contact point moves from one element to another, even if the consistent linearisation is used at the element level, the quadratic convergence of the Newton-Raphson method is affected. In more severe cases, when the percentage of the contact points changing the element is large, this influence can be so pronounced, that the iterative solution method becomes divergent. Such a danger exists in the case of beam-to-beam contact, where with the assumption of its point-wise character, the number of active constraints is relatively small. Hence, it is especially important to provide the C^1-continuity of curves representing beam edges and axes, i.e. the continuity of functions and its first derivatives should be ensured.

The bibliography related to smoothing of surfaces for 2D and 3D solids is quite broad. Numerous examples of geometry modeling were presented in the monograph by Farin (1993). Methods of smooth modeling of 2D solids in contact were dealt with, for instance, in the papers by: El Abbasi et al. (2001), where the third order spline functions were used, Stadler et al. (2003) with Non-Uniform Rational B-Splines (NURBS) or Wriggers and Krstulović (1999) with Bezier curves. Modeling of 3D surface geometry is more sophisticated but even in this range many effective methods were proposed. For instance, approaches with surfaces based on Bezier functions (Krstulović 2001, Krstulović et al. 2002), Gregory patches (Puso and Laursen 2002), smooth penetration function (Belytschko et al. 2002) or smooth function of normal forces in contact (Jones and Papadopoulos 2001) might be mentioned.

P. Litewka: Finite Element Analysis of Beam-to-Beam Contact, LNACM 53, pp. 71–97.
springerlink.com

In the majority of problems a smooth formulation of contact leads to very long formulae. It is especially true in the 3D analysis. Analytical calculation of tangent stiffness matrices for contact finite elements in these cases is impossible without resorting to computer programs. To this end the mathematical programs are used, which allow to carry out algebraic symbolic operations and to prepare ready-to-use codes in programming languages. One of such programs is SMS, based on the commercial mathematical program Mathematica, which was used in the paper by Krstulović et al. (2002).

A special attention can also be paid to mortar methods (Laursen 2002, Puso and Laursen 2004). They eliminate the necessity of smoothing because a modified definition of the penetration function is used. It is smeared on the length of contact segment or surface. Due to this, the constraints resulting from contact are not fulfilled in a point-wise way as in the classical slave-master approach but in an integral sense. Another virtue of these methods is also a better fulfillment of the non-penetration condition for finite element meshes of different sizes on both contacting bodies. However, these methods cannot be used in the beam-to-beam contact analysis due to its inherent point-wise character.

4.2 Smoothing of 3D Curves

4.2.1 General Remarks

In the analysis of contact between beams with rectangular cross-sections, presented in Chapter 2 shape functions in the form of the third-order polynomials are used to provide the C^1-continuity of curves representing edges of beams. These functions are taken directly from the formulation of the beam finite elements. In this chapter an alternative approach is suggested. The smoothing of curves used to represent axes of beams with circular cross-sections is carried out independently of the formulation of beam elements (Litewka 2007).

Here four methods of construction of a smooth curve are described. They involve two types of polynomial representation of a 3D curve, i.e. the Hermite's polynomials and the Bezier's curves, as well as two types of curve layout related to the beam nodes, i.e. the inscribed curve method and the node-preserving method.

4.2.2 Inscribed Curve Algorithm

In the case of the inscribed curve method a segment of a resulting C^1-continuous curve is constructed using three adjacent nodes, Fig. 4.1. The segment is spanned between the mid-points C_{12} and C_{23}, which are defined by the position vectors

$$\mathbf{x}_{12} = \frac{\mathbf{x}_2 + \mathbf{x}_1}{2} = (x_{12i})^T ,$$
$$\mathbf{x}_{23} = \frac{\mathbf{x}_3 + \mathbf{x}_2}{2} = (x_{23i})^T . \tag{4.1}$$

Furthermore, the curve is tangent to the straight lines passing through the nodes **1–2** and **2–3**, respectively. This ensures the C[1]-continuity between the adjacent segments. A similar concept was already successfully used in smoothing of 2D solids in contact (Wriggers and Krstulović 1999).

If the Hermite's polynomials are used to define the curve mathematically, then the curve can be represented parametrically using the dimensionless co-ordinate ξ, fulfilling the condition

$$-1 \le \xi \le 1$$

and each of three components of the position vector $\mathbf{x}$

$$\mathbf{x} = (x_1, x_2, x_3)^T \tag{4.2}$$

for an arbitrary point A on the curve can be expressed by

$$x_i = a_i \xi^3 + b_i \xi^2 + c_i \xi + d_i , \tag{4.3}$$

where $i = 1, 2$ or 3. In the vector form Eq. (4.3) can be written down as

$$x_i = (a_i, b_i, c_i, d_i)^T \circ (\xi^3, \xi^2, \xi, 1)^T = \mathbf{a}_i \circ \boldsymbol{\xi} . \tag{4.4}$$

To simplify the notation in the following the position vectors of the nodes **1**, **2** and **3** of the finite element mesh are represented by their components

$$\mathbf{x}_1 = (x_{11}, x_{12}, x_{13})^T = (x_{1i})^T ,$$
$$\mathbf{x}_2 = (x_{21}, x_{22}, x_{23})^T = (x_{2i})^T , \tag{4.5}$$
$$\mathbf{x}_3 = (x_{31}, x_{32}, x_{33})^T = (x_{3i})^T .$$

The unknown coefficients of the polynomials (4.3) are determined from the boundary conditions

$$\xi = -1 \;\Rightarrow\; x_i = x_{12i} = \frac{x_{2i} + x_{1i}}{2} ,$$
$$\xi = 1 \;\Rightarrow\; x_i = x_{23i} = \frac{x_{3i} + x_{2i}}{2} ,$$
$$\xi = -1 \;\Rightarrow\; \frac{\partial x_i}{\partial \xi} = \varphi_{12i} = \frac{x_{2i} - x_{1i}}{2 l_{12}} l_{123} , \tag{4.6}$$
$$\xi = 1 \;\Rightarrow\; \frac{\partial x_i}{\partial \xi} = \varphi_{23i} = \frac{x_{3i} - x_{2i}}{2 l_{23}} l_{123} ,$$

which force the curve to fulfil the C[1]-continuity conditions described above.

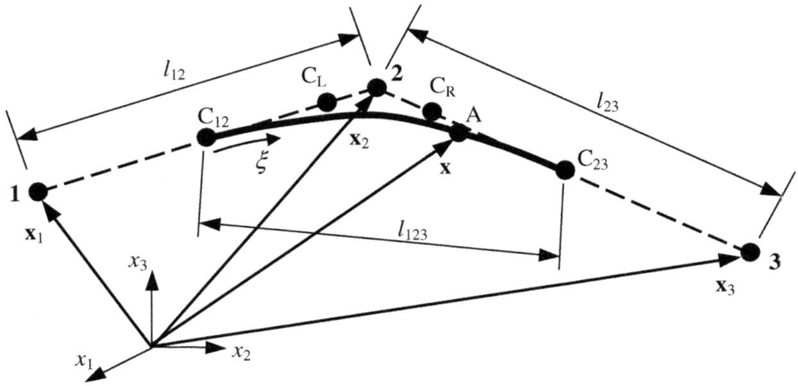

Fig. 4.1 Construction of inscribed curve segment

In $(4.6)_3$ and $(4.5)_4$, due to the relation between the dimensionless co-ordinate ξ and the co-ordinates x_1, x_2 and x_3, the straight line distances **1–2**, **2–3** and $\mathrm{C_{12}}$–$\mathrm{C_{23}}$ (Fig. 4.1) are present. They are denoted by l_{12}, l_{23}, and l_{123}, respectively, and can be calculated as

$$\begin{aligned} l_{12} &= \left\| \mathbf{x}_2 - \mathbf{x}_1 \right\|, \\ l_{23} &= \left\| \mathbf{x}_3 - \mathbf{x}_2 \right\|, \\ l_{123} &= \left\| \frac{\mathbf{x}_2 + \mathbf{x}_3}{2} - \frac{\mathbf{x}_1 + \mathbf{x}_2}{2} \right\|. \end{aligned} \tag{4.7}$$

Substituting the co-ordinates and their derivatives resulting from the approximation (4.4) to the boundary conditions (4.6) one gets for each of three co-ordinates, x_1, x_2 and x_3, a set of four equations with four unknown coefficients of the Hermite's polynomial. Solution of these sets can be expressed in terms of components of the position vectors and the slopes for the centre points

$$\mathbf{a}_i = \frac{1}{4}\begin{bmatrix} 1 & -1 & 1 & 1 \\ 0 & 0 & -1 & 1 \\ -3 & 3 & -1 & -1 \\ 2 & 2 & 1 & -1 \end{bmatrix} \begin{bmatrix} x_{c12i} \\ x_{c23i} \\ \varphi_{c12i} \\ \varphi_{c23i} \end{bmatrix} = \mathbf{A}_1 \begin{bmatrix} x_{c12i} \\ x_{c23i} \\ \varphi_{c12i} \\ \varphi_{c23i} \end{bmatrix} \tag{4.8}$$

or, finally, in terms of components of the position vectors for the element nodes

$$\mathbf{a}_i = \mathbf{A}_1 \frac{1}{2} \begin{bmatrix} 1 & 1 & 0 \\ 0 & 1 & 1 \\ -\frac{l_{123}}{l_{12}} & \frac{l_{123}}{l_{12}} & 0 \\ 0 & -\frac{l_{123}}{l_{23}} & \frac{l_{123}}{l_{23}} \end{bmatrix} \begin{bmatrix} x_{1i} \\ x_{2i} \\ x_{3i} \end{bmatrix} = \mathbf{A}_1 \, \mathbf{A}_2 \begin{bmatrix} x_{1i} \\ x_{2i} \\ x_{3i} \end{bmatrix} . \tag{4.9}$$

This vector of the coefficients of the Hermite's polynomial can be substituted to (4.4) to yield in the end the approximation of three components of the position vector for an arbitrary point on the smooth curve

$$x_i = \xi \circ \left[\mathbf{A}_1 \, \mathbf{A}_2 \left(x_{1i}, x_{2i}, x_{3i} \right)^T \right] . \tag{4.10}$$

It is worth to observe, that the relation (4.10) is non-linear with respect to the nodal position vectors, hence, also to the nodal displacements. Matrix $\mathbf{A}_2$ is not numeric but it depends non-linearly on these displacements via the lengths given in (4.7). In this situation, similarly as in the case of beams with rectangular cross-sections considered in Chapter 2, it is necessary to take into account the non-zero variables $\Delta\delta\mathbf{u}$ in the contact formulation.

On the other hand, the suggested approximation of the 3D curve depends only on the linear nodal displacements related to the global co-ordinate system and it is not a function of the nodal rotations. Thanks to this feature the presented beam-to-beam contact element with smoothing can be used in a combination with any, even the simplest possible beam elements. The fact, that the transformation from local to global co-ordinate system is not necessary is another issue contributing to the relative simplicity of the element.

In the further analysis expression of derivatives of the position vector components (4.10) with respect to the local co-ordinate ξ is necessary, too. They take the form

$$\begin{aligned} x_{i,\xi} &= \xi' \circ \left[\mathbf{A}_1 \, \mathbf{A}_2 \left(x_{1i}, x_{2i}, x_{3i} \right)^T \right] , \\ x_{i,\xi\xi} &= \xi'' \circ \left[\mathbf{A}_1 \, \mathbf{A}_2 \left(x_{1i}, x_{2i}, x_{3i} \right)^T \right] , \\ x_{i,\xi\xi\xi} &= \xi''' \circ \left[\mathbf{A}_1 \, \mathbf{A}_2 \left(x_{1i}, x_{2i}, x_{3i} \right)^T \right] \end{aligned} \tag{4.11}$$

with the following vectors

$$\begin{aligned} \xi' &= \left(3\xi^2, 2\xi, 1, 0 \right)^T , \\ \xi'' &= \left(6\xi, 2, 0, 0 \right)^T , \\ \xi''' &= \left(6, 0, 0, 0 \right)^T \end{aligned}$$

grouping the corresponding derivatives of the local co-ordinate.

The curve segment fulfilling the boundary conditions (4.6) ensuring the C^1-continuity of the entire curve can also be represented using the Bezier's curves. To this end it is necessary to construct a so called control polygon. It is proposed that the polygon is composed of the end nodes C_{12}, C_{23} and the intermediate nodes C_L, C_R. The latter nodes must lie on the straight lines **1–2** and **2–3**, respectively, see Fig. 4.1, to yield the curve tangent to these lines. Their position is somewhat arbitrary. Here a provision is taken to keep the polygon nodes at regular distances in order to get as regular curve as possible. It is adopted, that

$$\begin{aligned} \mathbf{x}_L &= \frac{5}{6}\mathbf{x}_2 + \frac{1}{6}\mathbf{x}_1 , \\ \mathbf{x}_R &= \frac{5}{6}\mathbf{x}_2 + \frac{1}{6}\mathbf{x}_3 . \end{aligned} \tag{4.12}$$

Then the Bezier's curve can be expressed in terms of Bernstein's polynomials (Farin 1993) to yield the co-ordinates (4.2) of the position vector $\mathbf{x}$ for a point A on the curve in the form

$$x_i = \frac{1}{8}\left[(1-\xi)^3 x_{12i} + 3(1+\xi)(1-\xi)^2 x_{Li} + 3(1+\xi)^2(1-\xi)x_{Ri} + (1+\xi)^3 x_{23i}\right]. \tag{4.13}$$

After substitution of (4.1) and (4.12) one can obtain the expression in terms of the nodal position vectors

$$x_i = \frac{1}{8}\left\{(1-\xi)^2 x_{1i} + 2\left[(1-\xi)^2(6+4\xi) + (1+\xi)^2(6-4\xi)\right]x_{2i} + (1+\xi)^2 x_{3i}\right\}. \tag{4.14}$$

This representation is very simple when compared to the one using the Hermite's polynomials (4.10). It is linear in the nodal displacements and the terms $\Delta\delta\mathbf{u}$ in the contact formulation vanish.

The derivatives of (4.14) with respect to the local co-ordinate are

$$\begin{aligned} x_{i,\xi} &= \frac{1}{4}\left[-(1-\xi)x_{1i} - 8\xi x_{2i} + (1+\xi)x_{3i}\right], \\ x_{i,\xi\xi} &= \frac{1}{4}\left(x_{1i} - 8x_{2i} + x_{3i}\right), \\ x_{i,\xi\xi\xi} &= 0 . \end{aligned} \tag{4.15}$$

4.2.3 Node-Preserving Algorithm

Arguably, a curve, which fits nodes better can be obtained, if a segment of the curve terminates at the nodes (Litewka 2008). The price to pay for this is a necessity to include four adjacent nodes in the analysis. The construction method for

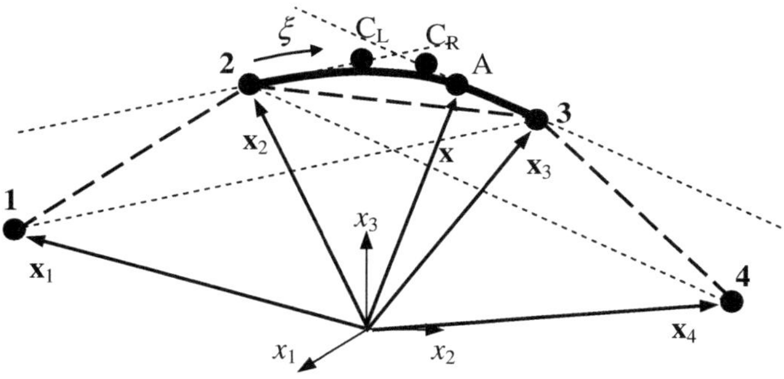

Fig. 4.2 Construction of node-preserving curve segment

this case is illustrated in Fig. 4.2. The segment ends coincide with the nodes **2** and **3** and the curve at these points is tangent to the lines parallel to the lines **1–3** and **2–4**, respectively. This ensures the C^1-continuity of the entire curve.

The boundary conditions necessary to find the coefficients of the Hermite's polynomial (4.3) expressing one component of the position vector $\mathbf{x}$ of a point A read

$$\begin{aligned}
&\xi=-1 \;\Rightarrow\; x_i = x_{2i}\,,\\
&\xi=1 \;\Rightarrow\; x_i = x_{3i}\,,\\
&\xi=-1 \;\Rightarrow\; \frac{\partial x_i}{\partial \xi} = \varphi_{13i} = \frac{x_{3i}-x_{1i}}{2l_{13}} l_{23}\,,\\
&\xi=1 \;\Rightarrow\; \frac{\partial x_i}{\partial \xi} = \varphi_{24i} = \frac{x_{4i}-x_{2i}}{2l_{24}} l_{23}\,,
\end{aligned} \tag{4.16}$$

where the position vector of the fourth node is represented similarly as those for the first three nodes (4.5)

$$\mathbf{x}_4 = \left(x_{41}, x_{42}, x_{43}\right)^T = \left(x_{4i}\right)^T. \tag{4.17}$$

Also, besides the straight line length l_{23} $(4.7)_2$, two other the lengths are defined

$$l_{13} = \left\|\mathbf{x}_3 - \mathbf{x}_1\right\|, \qquad (4.18)_1$$

$$l_{24} = \left\|\mathbf{x}_4 - \mathbf{x}_2\right\|. \qquad (4.18)_2$$

Again, after some algebra, similarly as in the case of the inscribed curve algorithm, the position vector $\mathbf{x}$ of a point A on the curve can be expressed in terms of the nodal position vectors in the following form

$$x_i = \begin{bmatrix}\xi^3\\ \xi^2\\ \xi\\ 1\end{bmatrix} \cdot \left\{ \frac{1}{8} \begin{bmatrix} 1 & -1 & 1 & 1\\ 0 & 0 & -1 & 1\\ -3 & 3 & -1 & -1\\ 2 & 2 & 1 & -1 \end{bmatrix} \begin{bmatrix} 0 & 2 & 0 & 0\\ 0 & 0 & 2 & 0\\ -l_{23}/l_{13} & 0 & l_{23}/l_{13} & 0\\ 0 & -l_{23}/l_{24} & 0 & l_{23}/l_{24} \end{bmatrix} \begin{bmatrix} x_{1i}\\ x_{2i}\\ x_{3i}\\ x_{4i}\end{bmatrix} \right\}. \tag{4.19}$$

This representation is reformulated introducing matrices

$$\mathbf{A}_3 = \frac{1}{8}\begin{bmatrix} 1 & -1 & 1 & 1 \\ 0 & 0 & -1 & 1 \\ -3 & 3 & -1 & -1 \\ 2 & 2 & 1 & -1 \end{bmatrix}, \quad \mathbf{A}_4 = \begin{bmatrix} 0 & 2 & 0 & 0 \\ 0 & 0 & 2 & 0 \\ -l_{23}/l_{13} & 0 & l_{23}/l_{13} & 0 \\ 0 & -l_{23}/l_{24} & 0 & l_{23}/l_{24} \end{bmatrix} \tag{4.20}$$

simplifying the notation to the form

$$x_i = \xi \circ \left[\mathbf{A}_3\ \mathbf{A}_4\ (x_{1i}, x_{2i}, x_{3i}, x_{4i})^T\right]. \tag{4.21}$$

Then, using the expressions similar to (4.11), the respective derivatives with respect to the local co-ordinate can be written down as

$$\begin{aligned} x_{i,\xi} &= \xi' \circ \left[\mathbf{A}_3\ \mathbf{A}_4\ (x_{1i}, x_{2i}, x_{3i}, x_{4i})^T\right], \\ x_{i,\xi\xi} &= \xi'' \circ \left[\mathbf{A}_3\ \mathbf{A}_4\ (x_{1i}, x_{2i}, x_{3i}, x_{4i})^T\right], \\ x_{i,\xi\xi\xi} &= \xi''' \circ \left[\mathbf{A}_3\ \mathbf{A}_4\ (x_{1i}, x_{2i}, x_{3i}, x_{4i})^T\right]. \end{aligned} \tag{4.22}$$

For the Bezier's representation of the node-preserving curve the control polygon must be constructed in a similar way as previously. The intermediate nodes C_L, C_R must lie on the lines parallel to the lines **1–3** and **2–4**, respectively, passing through the nodes **2** and **3**, respectively. Their position vectors are defined as

$$\begin{aligned} \mathbf{x}_L &= \frac{1}{3}(\mathbf{x}_3 - \mathbf{x}_1)\frac{l_{23}}{l_{13}} + \mathbf{x}_2, \\ \mathbf{x}_R &= -\frac{1}{3}(\mathbf{x}_4 - \mathbf{x}_2)\frac{l_{23}}{l_{24}} + \mathbf{x}_3, \end{aligned} \tag{4.23}$$

again with a purpose to obtain the curve segment as regular as possible.

The components of the position vector **x** can be expressed using the Bernstein's polynomials as

$$x_i = \frac{1}{8}\left[(1-\xi)^3 x_{2i} + 3(1+\xi)(1-\xi)^2 x_{Li} + 3(1+\xi)^2(1-\xi)x_{Ri} + (1+\xi)^3 x_{3i}\right] \tag{4.24}$$

or in terms of the nodal position vectors

$$x_i = \frac{1}{8}\left\{(1+\xi)(1-\xi)^2 \frac{l_{23}}{l_{13}} x_{1i} + \left[2(1-\xi)^2(2+\xi)+(1+\xi)^2(1-\xi)\frac{l_{23}}{l_{24}}\right]x_{2i} + \right.$$
$$\left. + \left[2(1+\xi)^2(2-\xi)+(1+\xi)(1-\xi)^2 \frac{l_{23}}{l_{13}}\right]x_{3i} + (1+\xi)^2(1-\xi)\frac{l_{23}}{l_{24}} x_{4i}\right\}. \tag{4.25}$$

With the relation (4.25) in hand the derivatives of this approximation with respect to the local co-ordinate can be found in the following form

$$x_{i,\xi} = \frac{1}{8}\left\{-(1-\xi)(1+3\xi)\frac{l_{23}}{l_{13}} x_{1i} + \left[6\left(\xi^2-1\right)+\left(1-2\xi+3\xi^2\right)\frac{l_{23}}{l_{24}}\right]x_{2i} + \right.$$
$$\left. + \left[-6\left(1+\xi^2\right)-\left(1+2\xi-3\xi^2\right)\frac{l_{23}}{l_{13}}\right]x_{3i} + (1+\xi)(1-3\xi)\frac{l_{23}}{l_{24}} x_{4i}\right\},$$
$$x_{i,\xi\xi} = \frac{1}{8}\left\{-(2-6\xi)\frac{l_{23}}{l_{13}} x_{1i} + \left[12\xi+(-2+6\xi)\frac{l_{23}}{l_{24}}\right]x_{2i} + \right. \tag{4.26}$$
$$\left. + \left[-12\xi-(2-6\xi)\frac{l_{23}}{l_{13}}\right]x_{3i} - (2+6\xi)\frac{l_{23}}{l_{24}} x_{4i}\right\},$$
$$x_{i,\xi\xi\xi} = \frac{1}{8}\left[6\frac{l_{23}}{l_{13}} x_{1i} + \left(12+6\frac{l_{23}}{l_{24}}\right)x_{2i} + \left(-12+6\frac{l_{23}}{l_{13}}\right)x_{3i} - 6\frac{l_{23}}{l_{24}} x_{4i}\right].$$

One can notice, that both curve representations (4.21) and (4.25), corresponding to the node-preserving algorithm, are non-linear in the nodal displacements, as is (4.10) for the Hermite's inscribed curve approximation.

4.3 Discretisation and Smooth Beam Contact Finite Elements

4.3.1 Inscribed Curve Elements

Application of the approximations (4.10) or (4.14) for the position vectors of points on the smooth curve, presented in Section 4.2.2, in the modeling of axes of two contacting beams with circular cross-sections leads to the formulation of the respective smooth beam contact finite elements. Each of these elements involves nodes of two pairs of adjacent beam finite elements, as shown in Fig. 4.3.

Implementation of such elements requires a certain modification of the contact search procedure presented in Section 2.3. In the first stage, instead of finding the pair of the closest elements it is more suitable to find a pair of the closest nodes. The further search of the contact points involves smooth segments of curves constructed on the base of the beam element pairs, to which the previously located closest nodes belong. Similarly as in the procedure described in Section 2.3, also

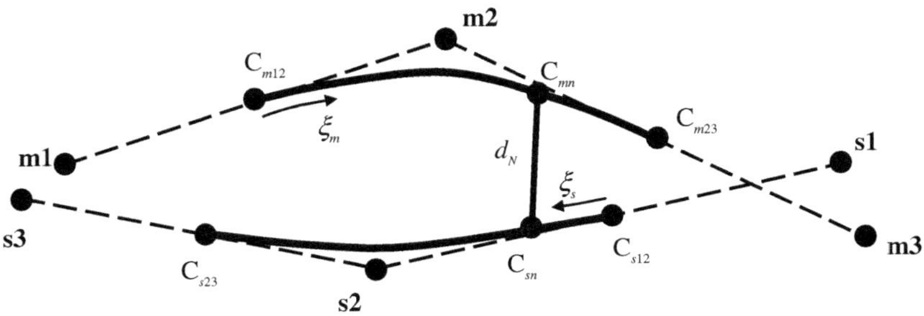

Fig. 4.3 Smooth beam contact finite element for inscribed curve approximation

here one must admit a possibility to check, if the local co-ordinates of the closest points C_{mn} and C_{sn}, lying on the smooth curves, yielding from orthogonality conditions (2.3), fulfil the constraints (2.14).

With the contact point candidates located one can calculate the penetration function (2.6) (the distance d_N is also depicted in Fig. 4.3) and check the contact criterion (2.7).

As was presented in Section 4.2 the proposed curve approximations do not depend on nodal cross-section rotations but only on linear displacements. Hence, the number of involved degrees of freedom is 18, i.e. three nodal displacements per each of six involved beam elements nodes, presented in Fig. 4.3. They can be assembled into the vector form

$$\mathbf{q} = \left(u_{m11}, u_{m12}, u_{m13}, u_{m21}, u_{m22}, u_{m23}, u_{m31}, u_{m32}, u_{m33}, \right.$$
$$\left. u_{s11}, u_{s12}, u_{s13}, u_{s21}, u_{s22}, u_{s23}, u_{s31}, u_{s32}, u_{s33}\right)^T = \left(\mathbf{u}_M{}^T, \mathbf{u}_S{}^T\right)^T . \tag{4.27}$$

Discretisation of the kinematic variables presented in Sections 2.5 and 3.4 is carried out in the same way for the presented smooth elemens. The same notation for the matrices can be applied and all the relations remain unchanged. The differences are in the way to determine the matrices $\mathbf{G}_{mn}$ and $\mathbf{G}_{sn}$ in (2.34) and (2.35), $\mathbf{H}_{mn}$ and $\mathbf{H}_{sn}$ in (2.36) and (2.37), $\mathbf{G}_{mp}$ and $\mathbf{G}_{sp}$ in (3.47) and (3.48) as well as $\mathbf{M}_{mn}$ and $\mathbf{M}_{sn}$ in (A2.11). Their components should be calculated as partial derivatives with respect to the nodal displacements $\mathbf{u}_M$ and $\mathbf{u}_S$ (4.12) of the non-linear approximation of displacements involving (4.10) or the linear one involving (4.14) as well as their derivatives (4.11) or (4.15). These calculations are carried out using the symbolic algebra program Maple 7 and the results of these calculations in a form of a ready-to-use Fortran code can be embedded in the computer program for the beam-to-beam contact analysis. Due to the very complex character of the approximations (4.10) and (4.14) the explicit representation of these matrices is too long to be presented here. Let it just be said, that the optimised Fortran code obtained in this way to calculate only the matrix $\mathbf{G}$ from (4.10) contains over 2300 lines. Hence, instead of presenting the matrices themselves, the exemplary

Maple 7 scripts without an output, which can be used to find these matrices for the case of Hermite's inscribed curve element are given in Appendix 3.

In the case of the inscribed curve representation each of these matrices has dimensions 3×9, what influences the dimensions of all further matrices included in the residual vectors and the tangent stiffness matrices of the contact finite elements. The matrix **F** and its components $\mathbf{F}_m$ and $\mathbf{F}_s$, used to calculate the linearisation and the variation of the local co-ordinates according to (2.39) and (2.40) have the dimensions 2×18 and 1×18, respectively. The auxiliary matrix **L** given in (2.41) is a 3×18 matrix.

The matrices $\mathbf{R}_m$ and $\mathbf{R}_s$ (A2.15), used in the calculation of the linearisation of the variations of the local co-ordinates according to (A2.10), have dimension 18×18. The same dimension characterizes all the matrices used to calculate the linearisation of the variation of the displacements for the current and previous contact points and their derivatives with respect to the local co-ordinate. The involved matrices are: $\mathbf{G}_{djmn}$ and $\mathbf{G}_{djsn}$ in (2.53), $\mathbf{G}_{djmp}$ and $\mathbf{G}_{djsp}$ in (3.51) as well as $\mathbf{H}_{djmn}$ and $\mathbf{H}_{djsn}$ in (A2.12), where $j = 1, 2$ or 3.

The change of dimension concerns also the matrices present in the friction formulation. The matrices $\mathbf{S}_m$ and $\mathbf{S}_s$, given in (3.52), are of dimension 3×18, and dimension 18×18 characterizes the matrices $\mathbf{P}_m$ and $\mathbf{P}_s$ in (3.54), $\mathbf{Z}_m$ and $\mathbf{Z}_s$ in (3.55) as well as $\mathbf{W}_m$ and $\mathbf{W}_s$ in (3.56).

Finally, the 18×18 tangent stiffness matrices and the 18-component residual vectors for both inscribed curve contact elements, based on Hermite's polynomials or Bezier's curves, can be determined using the appropriate formulae given in Sections 2.5 and 3.6. The matrices and vectors for friction vary depending on the friction state – stick or slip.

In all the calculations presented in Section 4.4 only the penalty method and the interpretation of the friction forces on the contacting beams as the components of one resultant friction force are used. So the following relations are applied: the stiffness matrix for the normal contact – (2.47), the stiffness matrix for the stick state – the sum of $(3.64)_1$ and $(3.64)_2$, the stiffness matrix for the slip state – the sum of $(3.68)_1$ and $(3.68)_2$, the residual vector for the normal contact – (2.44), the residual vector for the stick state – the sum of $(3.58)_1$ and $(3.58)_2$ and the residual vector for the slip state – the sum of $(3.62)_1$ and $(3.62)_2$.

Each active contact point adds to the global tangent stiffness matrix and the global residual vector the sum of the entries for the normal contact and the friction. The latter components depend on the current friction state – stick or slip.

4.3.2 Node-Preserving Elements

The node preserving smooth beam contact finite elements yield from the application of the approximations (4.21) or (4.25) for the position vector components of points on the smooth curve presented in Section 4.2.3 in the

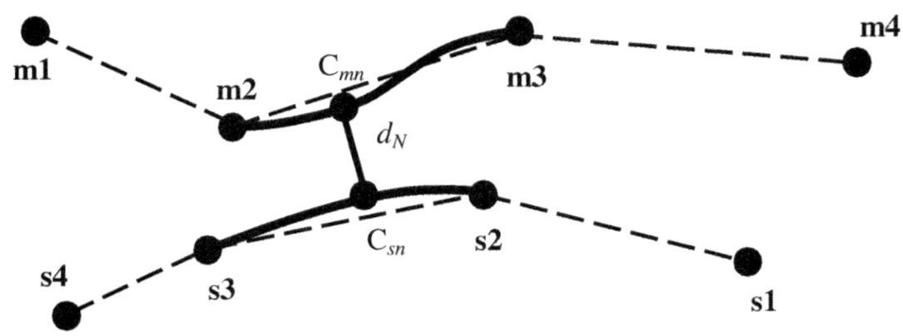

Fig. 4.4 Smooth beam contact finite element for node-preserving approximation

modeling of axes for two contacting beams with circular cross-sections. Each of these elements involves nodes of two sets of three adjacent beam finite elements, as shown in Fig. 4.4.

To these two contact elements one can apply the same first stage of the contact search as is described in Section 2.3. The closest elements found are the middle elements of the three element sets involved in the present formulation. The further search of the contact points involves smooth segments of the curves constructed on the base of three beam element sets, to which the previously located closest elements belong. Again, the located pair of the closest points must fulfil the constraints (2.14).

The number of involved degrees of freedom for the node-preserving elements, contrary to the inscribed curve elements, is 24, i.e. three nodal displacements per each of eight involved beam elements nodes, presented in Fig. 4.4. They can be assembled into the vector form

$$\begin{aligned}\mathbf{q} &= \left(u_{m11}, u_{m12}, u_{m13}, u_{m21}, u_{m22}, u_{m23}, u_{m31}, u_{m32}, u_{m33}, u_{m41}, u_{m42}, u_{m43},\right.\\ &\quad \left. u_{s11}, u_{s12}, u_{s13}, u_{s21}, u_{s22}, u_{s23}, u_{s31}, u_{s32}, u_{s33}, u_{s41}, u_{s42}, u_{s43}\right)^T \\ &= \left(\mathbf{u}_M{}^T, \mathbf{u}_S{}^T\right)^T .\end{aligned} \tag{4.28}$$

Since the dimension of the vector of degrees-of-freedom for these elements is the same as for the contact elements for beams with rectangular cross-sections presented in Chapters 2 and 3, all the notation used there can be directly used without any change of matrices dimensions. The only difference will be the method of calculation of the matrices $\mathbf{G}_{mn}$ and $\mathbf{G}_{sn}$ in (2.34) and (2.35), $\mathbf{H}_{mn}$ and $\mathbf{H}_{sn}$ in (2.36) and (2.37), $\mathbf{G}_{mp}$ and $\mathbf{G}_{sp}$ in (3.47) and (3.48) as well as $\mathbf{M}_{mn}$ and $\mathbf{M}_{sn}$ in (A2.11). Their components are found using the symbolic algebra program Maple 7 as was done for the inscribed curve elements in Section 4.3.1. The methods of finding the final forms of the tangent stiffness matrices and the residual vectors was also described there.

4.4 Numerical Examples

4.4.1 Introduction

There are two main purposes of the presented six numerical examples. In the first two cases, the calculations are performed using four types of the smooth beam contact finite elements. On the base of several criteria the most suitable element is chosen for the application in the four following examples, where this element is tested further with respect to, arguably, the most important criterion, i.e. the ability to perform effectively in the case of large sliding. Such situations pose crucial tests to check, if the suggested procedure of smoothing and the resulting contact elements meet the requirements.

In all the examples beams with circular cross-sections are analyzed. Similarly as in the examples presented in Chapters 2 and 3, each beam is discretized using 10 identical elements. The co-rotational beam element for large strains derived by Crisfield (1990) is chosen for the analysis. Crisfield stated in his paper, that the element led to an "almost" quadratic convergence of the Newton method, so the convergence of the entire contact problem with friction is also almost quadratic.

In all the presented examples beams are subjected to imposed nodal displacements applied simultaneously in a given number of equal increments. The selected moments of the deformation process are defined by the parameter T, varying in the range from 0 to 1. Iterations are continued until the relative energy, calculated as a scalar product of the current residual vector and the displacement increment vector, with respect to such an energy in the first iteration in a given increment is smaller than 10^{-20}.

The values of the penalty parameters used in the examples are selected using the trial-and-error method. The presented results are obtained with the maximal possible values, for which the convergence of the Newton-Raphson method was achieved.

Data for the calculations are given without any specified physical units, though it is understood, that any consistent unit set might be applied.

The presented graphical representations are restricted to beams axes only. To enhance the 3D views in the figures, axes of the beams "behind" are depicted as "broken".

Before the comparison of the presented four elements in the direct application to the contact problems, it is also interesting to compare the respective lengths of computer codes. The basic difference between all the formulations presented is the method of calculation of the matrices **G**, **H** and **M**, as discussed in Section 4.3. The length of the automatically generated optimized Fortran code for the considered four contact elements is given in Table 4.1.

It can be noted that the lack of non-linearity in the smooth curve representation for the inscribed curve method with Bezier's curves results in an extremely short code. So, at this stage, one might expect that the corresponding beam contact finite element would be the cheapest in terms of computer time.

Table 4.1 Number of lines in Fortran code for basic matrices in different smooth beam contact finite element formulations

Element type	Matrix **G**	Matrix **H**	Matrix **M**
Inscribed curve, Hermite	2291	1416	703
Inscribed curve, Bezier	51	45	27
Node-preserving, Hermite	983	959	838
Node-preserving, Bezier	718	764	652

4.4.2 Example 1

In this example contact between two cantilever beams shown in Fig. 4.5 is considered. The beams have circular cross-sections with radius $R = 0.1$, length $L = 6.0$ and are initially spaced at 0.001. They are made of a material with $E = 250{\cdot}10^5$, $\nu = 0.3$ and the friction coefficient $\mu = 0.8$. The penalty parameters used in the analysis are $\varepsilon_N = 1{\cdot}10^4$ and $\varepsilon_T = 1{\cdot}10^4$. The imposed displacements shown in Fig. 4.5 are applied in 60 equal increments. Displacements Δ_2= 1.0 and Δ_3= 0.5 are

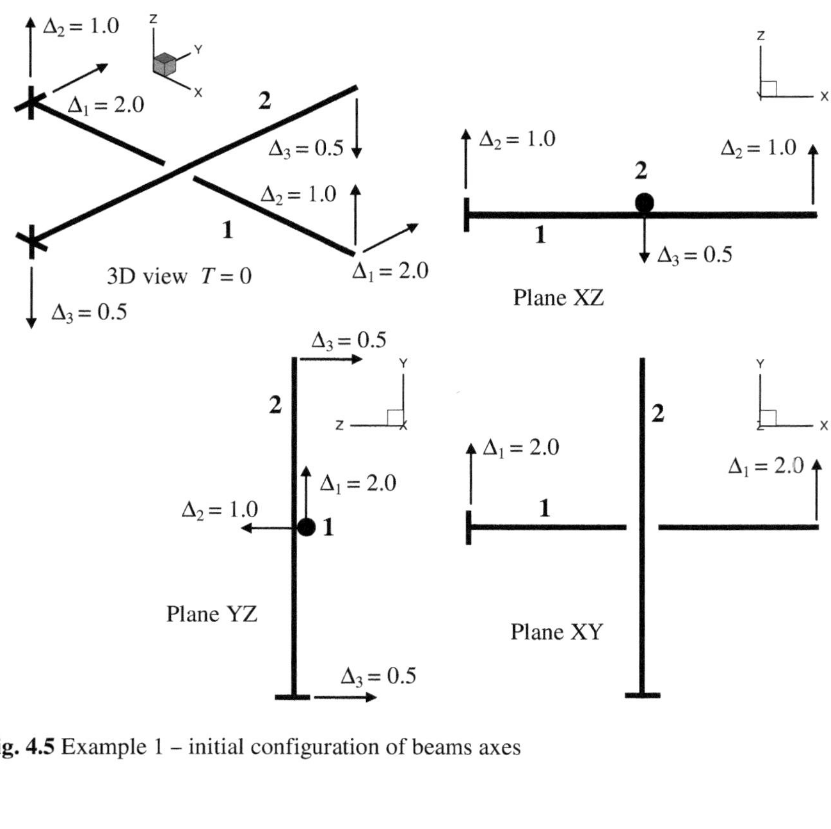

Fig. 4.5 Example 1 – initial configuration of beams axes

pressing the beams one against another leading them to contact, while displacements $\Delta_1 = 2.0$ cause the sliding of beam **1** along beam **2**.

In Fig. 4.6 the deformed layout of the beams axes is shown for the case of inscribed curve, Hermite's element. The results corresponding to the other elements would be impossible to distinguish.

The graphs presented in Fig. 4.7 show the evolution of the elastic parts of the tangential displacements for both beams. These variables are used, according to formulae (3.3) to calculate the components of the friction force. The smooth character of the curves in these graphs indicates, that in the analysed situation of contact, where one beam slides along another and the contact point travels along several adjacent segments of the smoothed curve, the value of the friction force changes in time in a continuous manner. This proves, that all the analysed beam

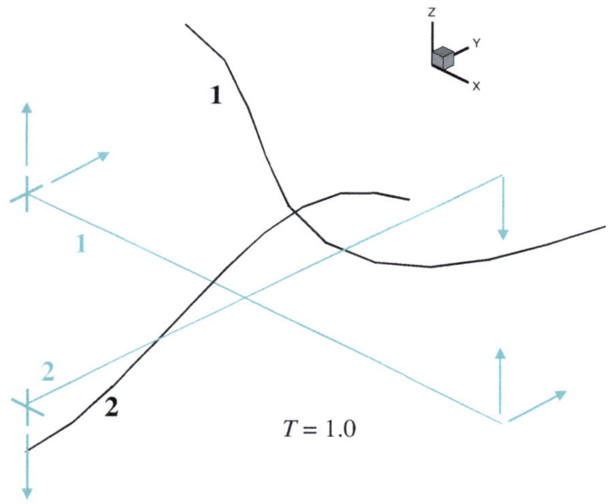

Fig. 4.6 Example 1 – deformed configuration of beams axes

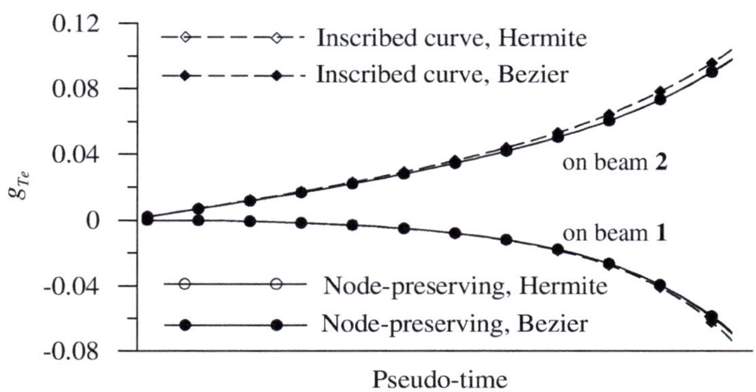

Fig. 4.7 Example 1 – evolution of elastic tangential gaps in deformation process

Table 4.2 Comparison of relative CPU time of analysis in Examples 1 and 2

Element type	Relative CPU time [%]	
	Example 1	Example 2
Inscribed curve, Hermite	100	100
Inscribed curve, Bezier	115	115
Node-preserving, Hermite	130	106
Node-preserving, Bezier	144	144

contact finite elements are efficient and that the main purpose of the smoothing procedure was achieved.

The differences between results for the inscribed curve and the node-preserving smoothing methods are present but very small. With the increasing number of beam elements used they are decreasing and for a reasonably large number the discrepancy is negligible. Actually, one might argue that 10 elements used here are enough to get the satisfactory accuracy. On the other hand, the influence of the curve representation using the Bezier's curves or Hermite's polynomials is negligible.

Table 4.2 presents the comparison of relative CPU times required for this analysis of the displacement process. It is very surprising to note, that the contact element with the simplest formulation, i.e. the inscribed curve Bezier does not yield the shortest time. On the other hand, the inscribed curve Hermite element with the longest code (see Table 4.1) gives the fastest results. The reason for this is a different number of iterations to the equilibrium state. It was found, that in almost all 60 increments used the inscribed curve Hermite element required at least one iteration less than the other elements. On the other hand, differences in the CPU time for a single increment using the considered elements are negligible.

4.4.3 Example 2

In this example contact between two beams shown in Fig. 4.8 is considered. The data for the beams are the same as in Example 1, except that beam **2** has now the length $L = 10.0$. Imposed displacements are applied in 60 increments. Displacements $\Delta_1 = 1.0$ and $\Delta_3 = 0.5$ are pressing the beams one against another, leading them to contact, while displacements $\Delta_2 = 3.0$ cause sliding of beam **2** along beam **1**.

In Fig. 4.9 the deformed layout of the beam axes is shown for the case of the inscribed curve Hermite element. As in Example 1, differences in the deformed shapes for the other contact elements are invisible.

The graphs presenting the evolution of the elastic gap, shown in Fig. 4.10, again prove the effectiveness of smoothing and the similarity of results obtained using the different analysed smoothing techniques.

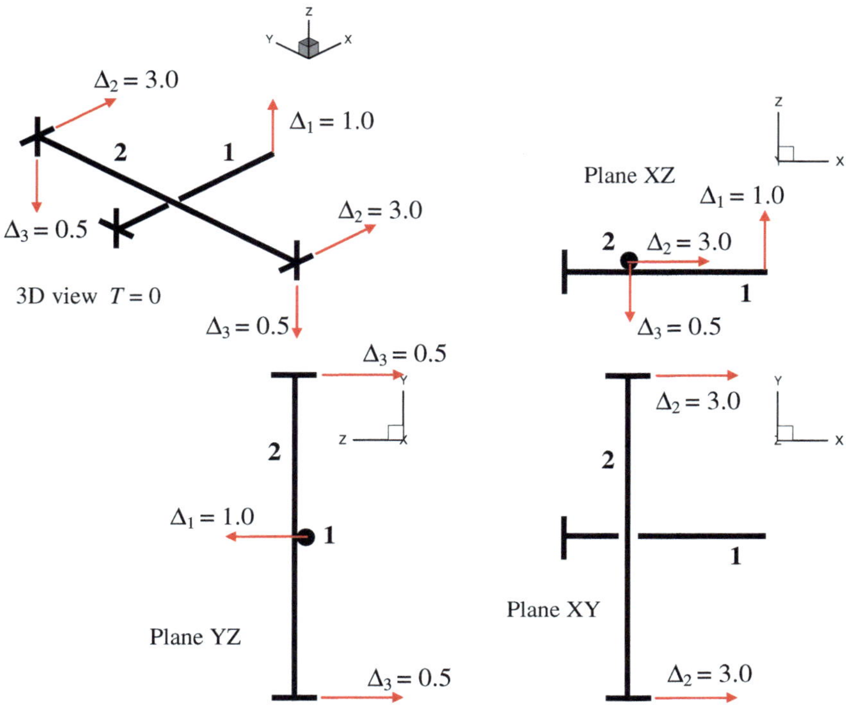

Fig. 4.8 Example 2 – initial configuration of beams axes

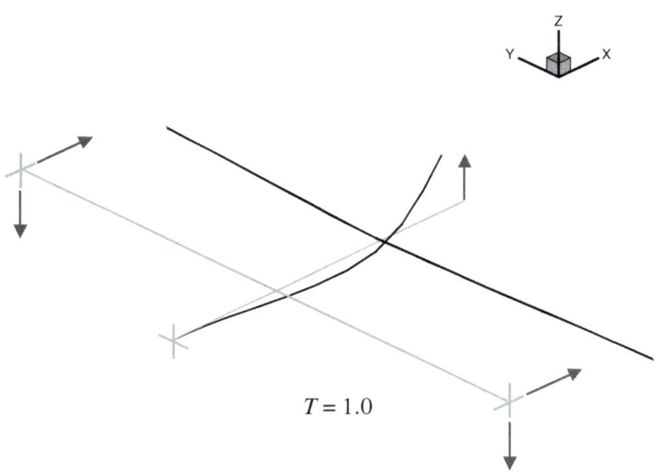

Fig. 4.9 Example 2 – deformed configuration of beams axes

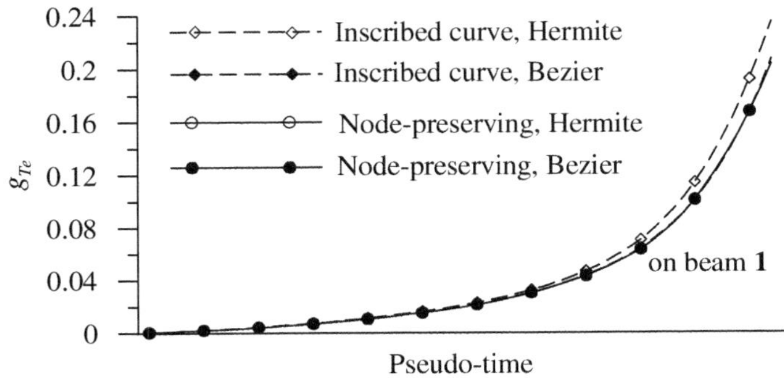

Fig. 4.10 Example 2 – evolution of elastic tangential gap in deformation process

Finally, the comparison of the CPU time required for the analysis, shown in the last column in Table 4.2 leads to similar conclusions as in Example 1. In the view of these results, the inscribed curve Hermite element can be considered as the best choice and this element is used in further analyses presented in the following examples.

4.4.4 Example 3

In this example contact between two cantilever beams, which axes in the initial configuration are presented in Fig. 4.11 is analysed. Imposed displacements $\Delta_1 = \Delta_2 = 0.5$ applied at the free ends lead the beams to contact and the displacement $\Delta_3 = 2.0$ applied at the support of beam **1** causes its sliding along beam **2**. The displacements are applied in 120 increments. The following data are used in the calculations: beam **1** – $E = 205{\cdot}10^5$, $\nu = 0.3$, radius of circular cross-section 0.1, length 6.0; beam **2** – $E = 70{\cdot}10^5$, $\nu = 0.25$, radius of circular cross-section 0.15, length 6.0; initial gap between the beams 0.01; penalty parameters $\varepsilon_N = 40000$, $\varepsilon_{Tm} = \varepsilon_{Ts} = 8000$; friction coefficient $\mu = 0.2$ or 0.5. The example is solved using the inscribed curve Hermite contact element.

Deformed configurations of the beams axes corresponding to four selected moments of the deformation process for both analysed values of the friction coefficient are presented in Fig. 4.12. The influence of these friction properties on the results is very clear. For $\mu = 0.5$ the stick state is present in the contact point and the relative movement of the beams is restrained. This leads to larger flexural deformations. On the other hand, for $\mu = 0.2$ the slip state occurs. This can also be observed in the graphs presented in Fig. 4.13, where the evolution of the plastic parts of the tangential displacements during the deformation process is shown.

The graphs presented in Fig. 4.14 show the evolution of the elastic parts of the tangential displacements for both beams. These variables are used, according to

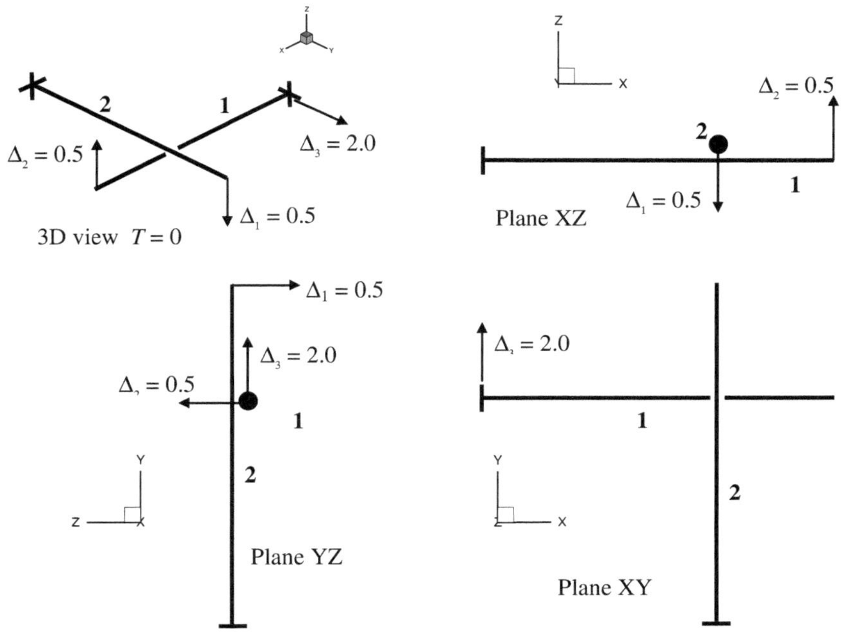

Fig. 4.11 Example 3 – initial configuration of beams axes

formulae (3.3), to calculate the components of the friction force. The smooth character of the curves in these graphs indicates, that in the analysed situation of contact, where one beam slides along another and the contact point travels along several adjacent segments of the smoothed curve, the value of the friction force

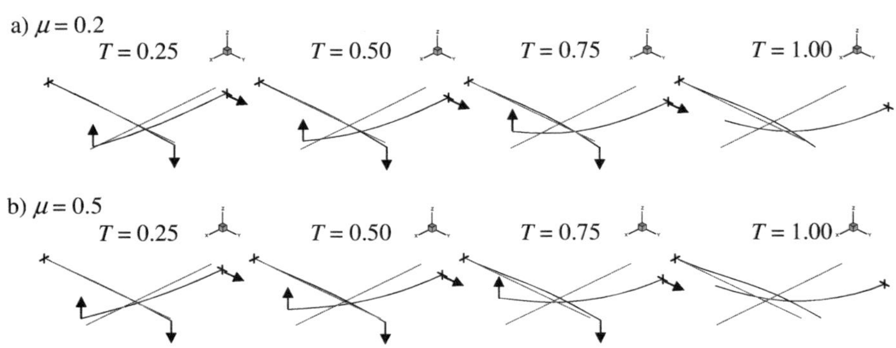

Fig. 4.12 Example 3 – deformed configurations of beams axes for: a) $\mu = 0.2$, b) $\mu = 0.5$

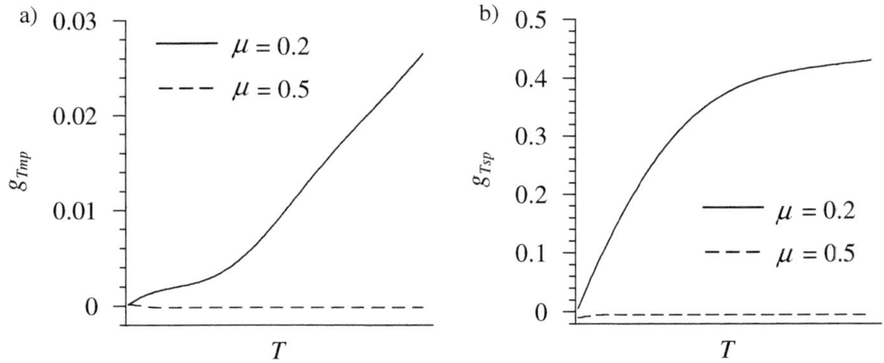

Fig. 4.13 Example 3 – evolution of plastic parts of tangential displacements on: a) beam **1**, b) beam **2**

changes in time in a continuous manner. This proves, that the analysed beam contact finite element is efficient and that the main purpose of the smoothing procedure was achieved.

Results of analysis of convergence in the iterative Newton-Raphson method are presented in Table 4.3. During the entire deformation process 5 to 6 iterations were required to achieve the equilibrium state. The convergence is not quadratic, especially at the end of the process. It results from the type of the applied beam finite element and also from the presence of very large contact forces at the end of the process, which influence the accuracy of numerical results. It is especially visible for the very small energy tolerance, which is used in the presented examples.

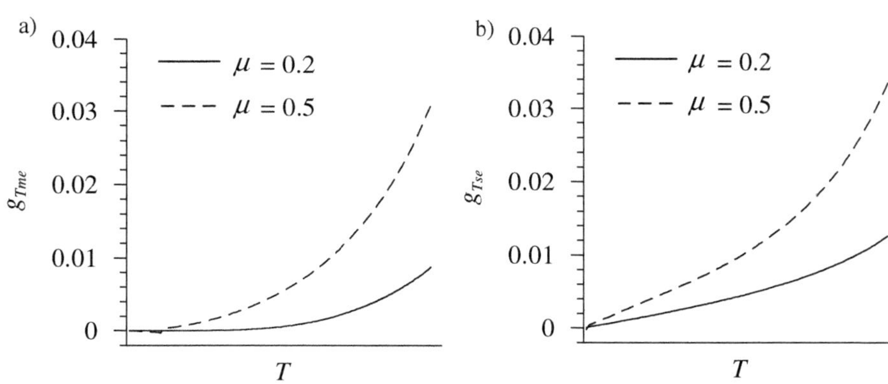

Fig. 4.14 Example 3 – evolution of elastic parts of tangential displacements on: a) beam **1**, b) beam **2**

Table 4.3 Example 3 – convergence analysis

Increment number		30	60	90	120
T		0.25	0.50	0.75	1.00
Relative energy for the iteration	1	1.0	1.0	1.0	1.0
	2	$3{,}8\times10^{-5}$	$4{,}1\times10^{-5}$	$4{,}6\times10^{-5}$	$5{,}1\times10^{-5}$
	3	$2{,}4\times10^{-13}$	$3{,}2\times10^{-12}$	$2{,}6\times10^{-11}$	$1{,}7\times10^{-10}$
	4	$1{,}1\times10^{-19}$	$1{,}1\times10^{-17}$	$6{,}3\times10^{-16}$	$3{,}0\times10^{-14}$
	5	$1{,}3\times10^{-25}$	$5{,}6\times10^{-23}$	$1{,}8\times10^{-20}$	$5{,}7\times10^{-18}$
	6	–	–	$5{,}4\times10^{-25}$	$1{,}1\times10^{-21}$

4.4.5 Example 4

In this example contact between beams, which axes in the initial configuration are presented in Fig. 4.15 is analysed. Beam **1** is clamped-clamped and beam **2** is a cantliever, which support is subjected to two displacements $\Delta_1 = 1.0$ and $\Delta_2 = 4.0$ causing respectively, contact and sliding between the beams. The displacements are applied in 30 increments. The following data are taken to the calculations: beam **1** – $E = 70{\cdot}10^5$, $\nu = 0.3$, radius of circular cross-section 0.1, length 6.0; beam **2** – $E = 205{\cdot}10^5$, $\nu = 0.3$, radius of circular cross-section 0.15, length 6.0;

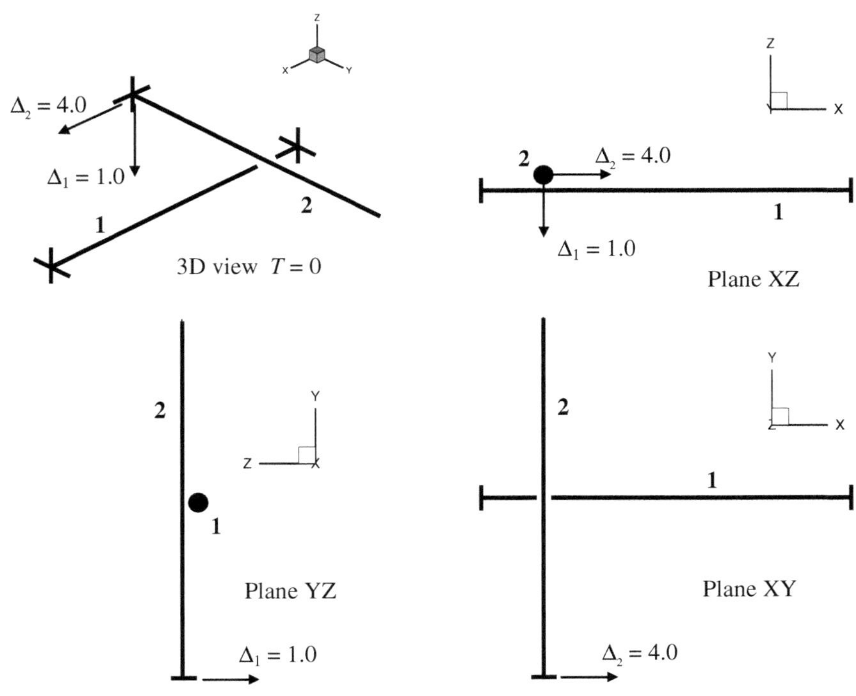

Fig. 4.15 Example 4 – initial configuration of beams axes

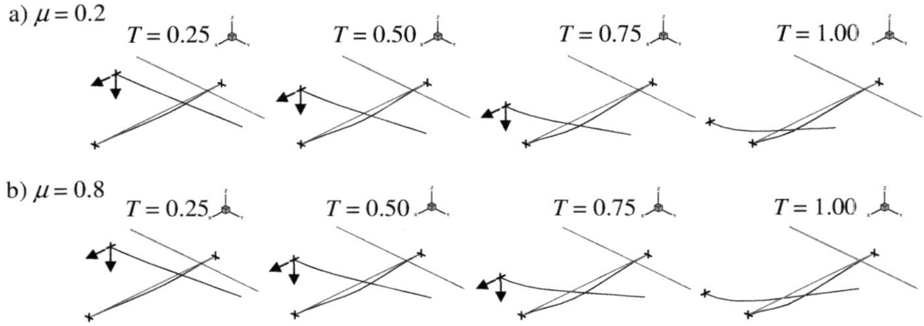

Fig. 4.16 Example 4 – deformed configurations of beams axes for: a) $\mu = 0.2$, b) $\mu = 0.8$

initial gap between the beams 0.001; penalty parameters $\varepsilon_N = 30000$, $\varepsilon_{Tm} = \varepsilon_{Ts} = 10000$; friction coefficient $\mu = 0.2$ or 0.8. The example is solved using the inscribed curve Hermite contact element.

Deformed configurations of the beams axes in four selected moments of the deformation process for both analysed values of the friction coefficient are presented in Fig. 4.16. In this example, due to the large value of displacement Δ_2 causing sliding, in both cases the slip state occurs. However, for the larger friction sliding is smaller, what leads to a larger flexural deformation of beam **2**.

Graphs presented in Fig. 4.17 show the evolution of the plastic parts of the tangential displacements during the deformation process. Contrary to Example 3, here these values are non-zero for both values of friction coefficient. It is also worth to note, that there is a change of sliding direction along beam **2**. In the initial

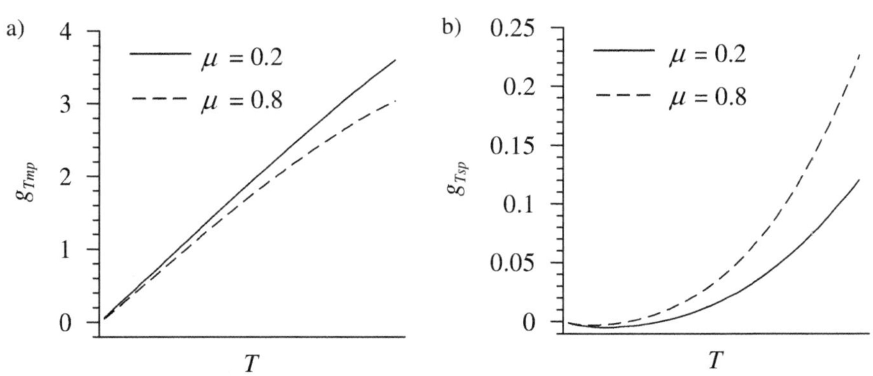

Fig. 4.17 Example 4 – evolution of plastic parts of tangential displacements on: a) beam **1**, b) beam **2**

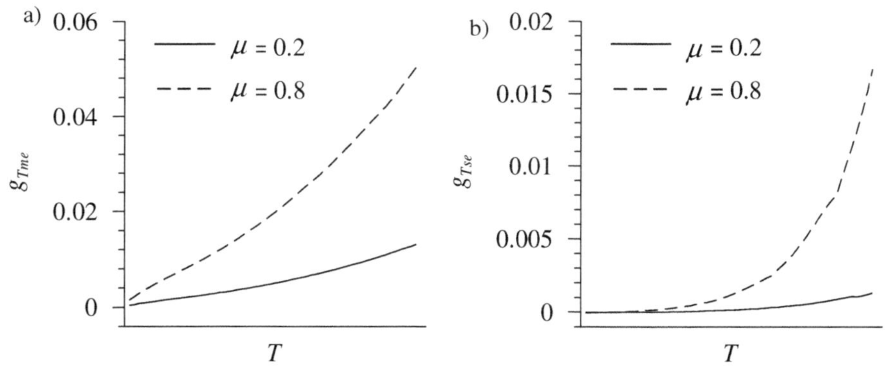

Fig. 4.18 Example 4 – evolution of elastic parts of tangential displacements on: a) beam **1**, b) beam **2**

phase a small negative plastic displacement takes place, what can be observed in Fig. 4.17b.

In this example the contact point moves along six subsequent finite elements on beam **2**. Without the procedure of curve smoothing the normal and tangent vectors and in consequence – values of the forces at the contact point, would exhibit discontinuities. Hence, the effectiveness of the suggested smooth contact finite element is confirmed again by the graphs presented in Fig. 4.18. They show the smooth character of the evolution of the elastic part of the tangential displacements and, indirectly – the friction force, during the deformation process.

In this example 5 to 7 iterations per increment were necessary to achieve the state of equilibrium.

4.4.6 Example 5

In this example contact between three cantilever beams beams, which axes in the initial configuration are presented in Fig. 4.19 is analysed. At the free end of beam **3** displacement $\Delta = 1.0$ is applied. This displacement, applied in 30 increments, causes a domino effect and leads to the mutual contacts between three beams. The following data are taken to the calculations: beam **1** – $E = 205{\cdot}10^5$, $\nu = 0.3$, radius of circular cross-section 0.1, length 6.0; beam **2** – $E = 200{\cdot}10^5$, $\nu = 0.25$, radius of circular cross-section 0.15, length 6.0; beam **3** – $E = 70{\cdot}10^5$, $\nu = 0.25$, radius of circular cross-section 0.15, length 6.0; initial gaps between beams **1** and **2** as well as **2** and **3** 0.01; penalty parameters $\varepsilon_N = 20000$, $\varepsilon_{Tm} = \varepsilon_{Ts} = 4000$; friction coefficient $\mu = 0.0$ or 0.478. The example is solved using the inscribed curve Hermite contact element.

Deformed configurations of the beams axes for both cases of the friction coefficient in selected four moments of the deformation process are presented in Fig. 4.20. In this example two points of contact exist. For $\mu = 0.0$ both are in the

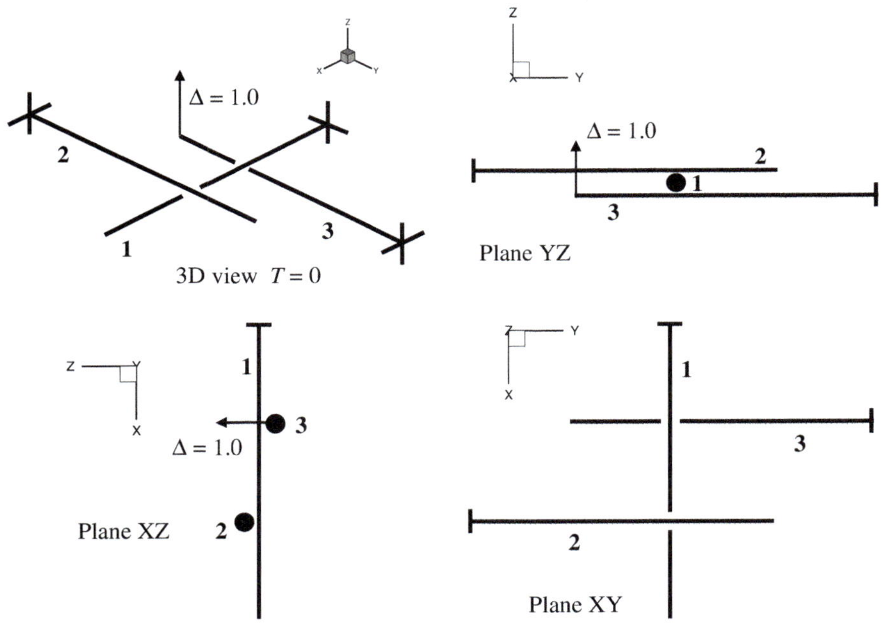

Fig. 4.19 Example 5 – initial configuration of beams axes

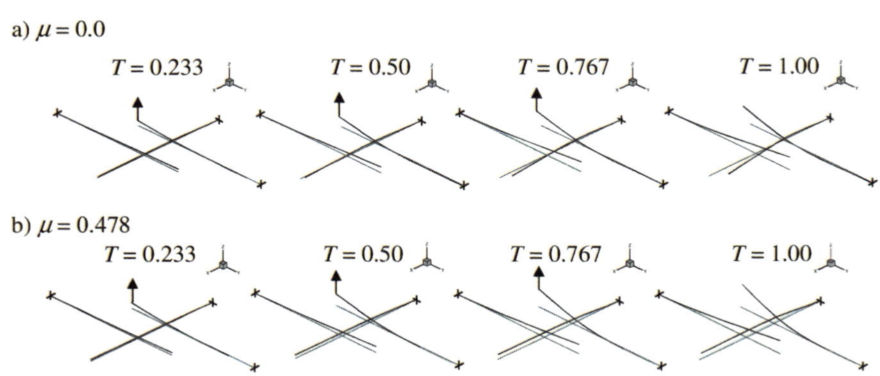

Fig. 4.20 Example 5 – deformed configurations of beams axes for: a) $\mu = 0.0$, b) $\mu = 0.478$

slip state. On the other hand, for $\mu = 0.478$ contact between beams **1** and **3** is characterised by the stick state during the entire process. For the contact between beams **1** and **2** a change from stick to slip takes place at $T = 0.6$. This phenomenon can be observed in the graphs shown in Fig. 4.21, where the evolution of the plastic parts of the tangential displacements during the process is presented. In this case sliding occurs mostly along beam **2** with the relative displacement about 10

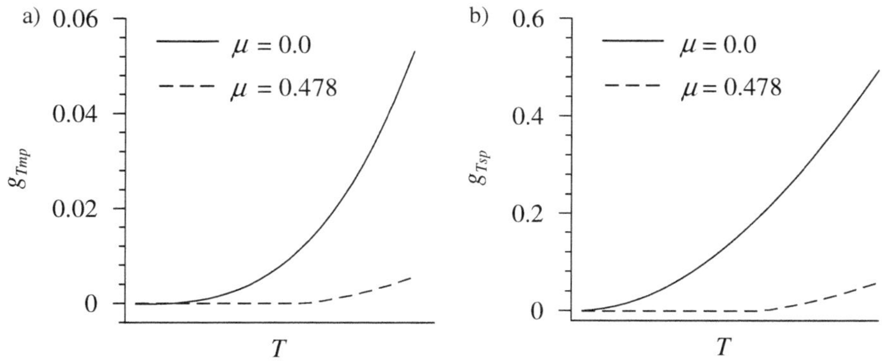

Fig. 4.21 Example 5 – evolution of plastic parts of tangential displacements on: a) beam **1**, b) beam **2**

times larger than the corresponding value for beam **1**. It can also be seen in Fig. 4.20, if one compares the displacements of tips of both beams. Also, the influence of the friction and, in consequence – the friction state in contact **1** – **2**, on the displacement of the beam **3** tip in direction of the beam **1** axis (X) is notable.

In the analysed example 4 to 6 iterations per increment were necessary to achieve the equilibrium state.

The purpose of this example was to check the algorithm also for multiple contact points. Large sliding between facets did not occur here and the respective graphs of the tangential displacements are continuous.

4.4.7 Example 6

In this example contact between four identical beams, forming a symmetric assembly is analysed. The beams axes in the initial configuration are presented in Fig. 4.22. The beams have almost fully constrained centre points, except for the freedom of rotation about the axes lying in the plane XY perpendicular to the beams. At the free ends equal displacements of $\Delta = 1.5$ are applied in 30 increments. These displacements lead to simultaneous contact between the beams. The following data are taken to the calculations: $E = 205 \cdot 10^5$, $\nu = 0.3$, radius of circular cross-section 0.1, length 8.01; initial gaps between the beams 0.019; penalty parameters $\varepsilon_N = 40000$, $\varepsilon_{Tm} = \varepsilon_{Ts} = 7000$; friction coefficient $\mu = 0.0$ or 0.8. The example is solved using the inscribed curve Hermite contact element.

Deformed configurations of the beams axes for four selected moments of the deformation process for the both analysed friction cases are presented in Fig. 4.23. In this example four contact points are present but due to the symmetry displacements and forces therein are identical.

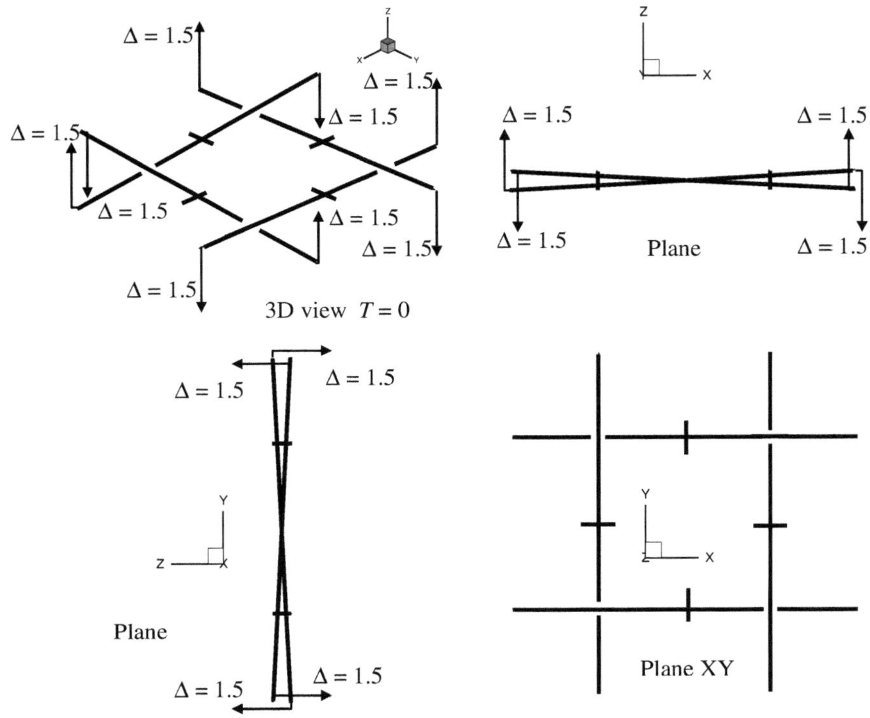

Fig. 4.22 Example 6 – initial configuration of beams axes

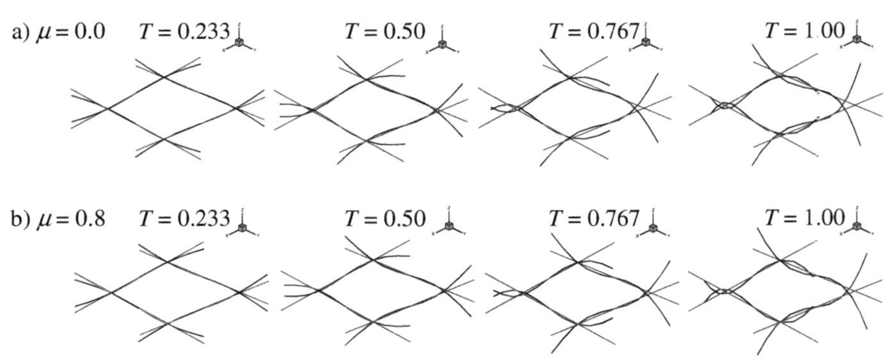

Fig. 4.23 Example 6 – deformed configurations of beams axes for: a) $\mu = 0.0$, b) $\mu = 0.8$

The graph in Fig. 4.24 presents the development of the total tangential displacement in the contact. For $\mu = 0.0$ the elastic part of the displacement is zero, so the solid line in the graph depicts its plastic part. On the other hand, for $\mu = 0.478$, which case is characterised by the stick state, the plastic part is zero, so the dashed line depicts the elastic part of the displacement. The exact solution would require this elastic part to vanish. However, the adopted values of the penalty parameters cannot be increased further to decrease these displacements due to the ill-posedness of the problem and difficulties with convergence of the Newton method. However, one might argue that the achieved accuracy is good enough because the above mentioned elastic part of the displacement is equal to 0.04, what is just 1/5 of the beam cross-section diameter and 1/200 of its length.

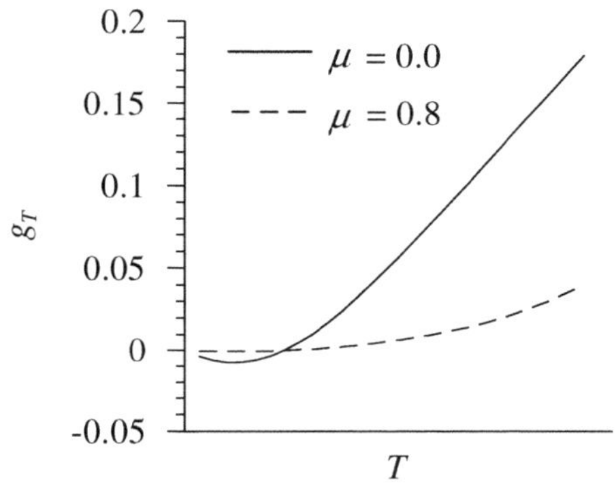

Fig. 4.24 Example 6 – evolution of tangential displacements

In the analysed example 5 to 6 iterations per increment were necessary to achieve the equilibrium state.

Chapter 5
Electric Contact

5.1 Introduction

The purely mechanical contact forumulation presented in the previous chapters may not suffice in cases, when complicated physical phenomena occuring in contact in advanced fields of engineering must be taken into account. Nuclear or space engineering as well as mechanics of electric conductors or superconductors are the representative examples. In such a situation it is necessary to consider a coupling between the mechanical fields of displacements or strains and electric, magnetic or thermal fields in the contacting objects.

Analysis of contact between electric conductors, due to their shape, allows to apply the beam-to-beam contact formulation modified with additional effects. One of the simplest possibilities to treat such problems is a consideration of a mechanical-electric coupling. One of the first attempts to solve it can be found in the paper by Boso et al. (2005a), devoted to contact between 2D solids.

In this chapter electro-mechanical beam-to-beam contact formulation (Boso et al. 2005b) is presented, which can be used in the analysis of electric conductors. The considerations are limited to the flow of the direct electric current. Influences of forces due to the resulting magnetic field are neglected an so are the variations of physical properties of materials with temperature as well as heat production due to the electric current flow. With these assumptions the formulation can be restricted to the coupling between the frictionless contact and the electric field. The friction can be added independently. Besides, the contact is assumed to be pointwise, between uniform, physically clean and smooth surfaces of electric conductors with circular cross-sections.

Electric phenomena taking place at the contact of two electric conductors were analysed and described in the monograph by Holm (1981). Basing on those considerations it can be concluded, that the contact of two long conductors occurs at a small area. In such a case one is allowed to assume a so called long constriction in the unconstrained flow of the electric current. The value of electric resistance for such a feature can be easily found.

5.2 Electro-mechanical Variables for Contact

The purely mechanical aspects of frictionless contact between beams were discussed in Chapter 2. In order to include additionally the electric field, new

P. Litewka: Finite Element Analysis of Beam-to-Beam Contact, LNACM 53, pp. 99–119.
springerlink.com

variables must be incorporated. For the static electric field the corresponding variable is the voltage. Let us assume, that in a given cross-section of a beam the voltage is given by the value $V(\xi)$. Then for a pair of contact points C_{mn} and C_{ms}, (Fig. 2.2) found in the contact search procedure discussed in Section 2.3, a voltage difference or a voltage gap can be introduced

$$g_V = V_{mn} - V_{sn}\,, \tag{5.1}$$

where V_{mn} and V_{sn} denote voltages in the cross-sections of the beams m and s, respectively, corresponding to the contact points. Their values can be expressed in terms of the local co-ordinates for these points

$$\begin{aligned} V_{mn} &= V_m\left(\xi_m = \xi_{mn}\right), \\ V_{sn} &= V_s\left(\xi_s = \xi_{sn}\right). \end{aligned} \tag{5.2}$$

According to the Ohm's law the electric current flowing through the contact between two conductors can be calculated as

$$I = g_V h_V\,, \tag{5.3}$$

where h_V denotes the electric conductivity of the contact.

With an assumption, that the length of the conductors is much larger than the diameter of the cross-sections and that the contact area is much smaller than the cross-section itself, the long constriction can be considered (Holm 1981) and its electric conductivity is given by

$$h_V = 2K_{eq}a\,, \tag{5.4}$$

where a is the radius of the circular contact area. The quantity K_{eq} in Eq. (5.4) is the mean electric conductivity determined as

$$K_{eq} = \frac{K_m K_s}{K_m + K_s} \tag{5.5}$$

using the values of the electric conductivity, K_m and K_s, for the materials composing both conductors.

In a general case contact between two electric conductors with circular cross-sections can be characterised by an elliptic contact area, as shown in Fig. 5.1. In such a situation to be able to use the relation (5.4) one can introduce the equivalent radius a determined as a geometric mean value of the semi-axes of the resulting ellipse, denoted by c and d

$$a = \sqrt{cd}\,. \tag{5.6}$$

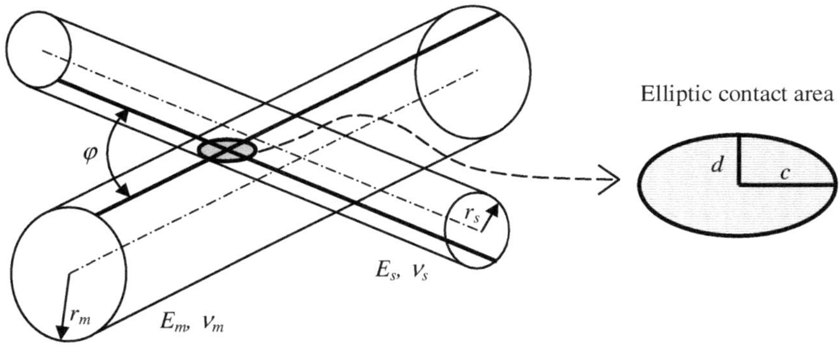

Fig. 5.1 Elliptic contact area for two contacting cylinders

For the sake of simplicity, results of the Hertz's theory of contact between ideally elastic bodies are used to calculate the lengths of the semi-axes of the ellipse. One can use the handbook by Young and Budynas (2001) or the monograph by Johnson (1985), where the corresponding formulae for different cases of geometry of contacting bodies are presented. For two contacting cylinders with radii of circular cross-sections r_m and r_s, made of materials with Young moduli, E_m and E_s, and Poisson's ratios, ν_m and ν_s, which axes form an angle φ in the plane, the corresponding lengths of the semi-axes of the elliptic contact area are given by

$$c = \alpha \sqrt[3]{\frac{3F_N r_{eq}}{2E_{eq}}}, \qquad d = \beta \sqrt[3]{\frac{3F_N r_{eq}}{2E_{eq}}}, \tag{5.7}$$

where the mean values of radius of circular cross-sections

$$r_{eq} = \frac{r_m r_s}{r_m + r_s} \tag{5.8}$$

and of Young's modulus

$$E_{eq} = \frac{E_m E_s}{E_m\left(1-\nu_s^{\,2}\right) + E_s\left(1-\nu_m^{\,2}\right)} \tag{5.9}$$

are introduced. Coefficients α and β in Eq. (5.7), depending on the radii of the circular cross-sections and the angle φ are tabelarised in the handbook by Young and Budynas (2001). Their values can also be calculated from the simplified interpolation formulae

$$\alpha = 1 - \ln(1-x) - 0{,}32x + 3{,}35x^{40}\,, \qquad \beta = 1 - 0{,}57x - 0{,}19x^{20}\,, \tag{5.10}$$

where the parameter x is given by the relation

$$x = r_{eq}\sqrt{\frac{1}{r_m^{\;2}} + \frac{1}{r_s^{\;2}} + \frac{2}{r_m r_s}\cos 2\varphi}\,. \tag{5.11}$$

After substitution of (5.7) to (5.6) the equivalent radius of the circular contact area can be expressed as a function of the normal contact force F_N as

$$a = \sqrt{\alpha\beta}\;\sqrt[3]{\frac{3F_N r_{eq}}{2E_{eq}}}\,. \tag{5.12}$$

This value substituted to (5.4) allows to write down the relation between the conductivity in the long constriction and the normal force in the contact

$$h_V = 2K_{eq}\sqrt{\alpha\beta}\;\sqrt[3]{\frac{3F_N r_{eq}}{2E_{eq}}}\,. \tag{5.13}$$

If the penalty method is used in the contact analysis, then the normal contact force can be expressed as

$$F_N = \varepsilon_N g_N\,. \tag{5.14}$$

Hence, finally, after taking advantage of the Ohm's law (5.3), the electric current in the contact can be related to the variables of electro-mechanical contact – the penetration function g_N and the voltage gap g_V

$$I = 2K_{eq}\sqrt{\alpha\beta}\;\sqrt[3]{\frac{3\varepsilon_N r_{eq}}{2E_{eq}}}\;\sqrt[3]{g_N}\;g_V\,. \tag{5.15}$$

In the following considerations partial derivatives of the current (5.15) with respect to the contact variables g_N and g_V are necessary. They are

$$\frac{\partial I}{\partial g_N} = \frac{2}{3}K_{eq}\sqrt{\alpha\beta}\;\sqrt[3]{\frac{3\varepsilon_N r_{eq}}{2E_{eq}}}\;\frac{1}{\sqrt[3]{g_N^{\;2}}}\,g_V\,, \qquad \frac{\partial I}{\partial g_V} = 2K_{eq}\sqrt{\alpha\beta}\;\sqrt[3]{\frac{3\varepsilon_N r_{eq}}{2E_{eq}}}\;\sqrt[3]{g_N}\,. \tag{5.16}$$

Besides, in the formulation of the electro-mechanical contact the kinematic variables calculated from the penetration function g_N and the voltage gap g_V are present. In order to distinguish between the variations and the linearisations calculated with respect to the mechanical variables (displacements) and the electric variables (voltages) subscripts u and V, respectively, are introcuced at the corresponding symbols δ and Δ. In this new notation the variables calculated from the penetration function g_N with respect to displacements, given by (2.25), are now denoted as $\Delta_u g_N$, $\delta_u g_N$ and $\Delta_u \delta_u g_N$. On the other hand, the variables calculated as the variations and the linearisations of g_N with respect to voltage are zero

$$\begin{aligned} \Delta_V g_N &= 0, \\ \Delta_V \delta_u g_N &= 0 \end{aligned} \tag{5.17}$$

because within the adopted assumptions the value of g_N does not depend on voltage.

The variables calculated from the function g_V with respect to displacements are non-zero

$$\begin{aligned} \Delta_u g_V &= \Delta_u (V_{mn} - V_{sn}) = V_{mn,m} \Delta_u \xi_{mn} - V_{sn,s} \Delta_u \xi_{sn}, \\ \Delta_u \delta_V g_V &= \delta_V (\Delta_u g_V) = \delta_V V_{mn,m} \Delta_u \xi_{mn} - \delta_V V_{sn,s} \Delta_u \xi_{sn} \end{aligned} \tag{5.18}$$

because this function depends on location of the contact points, which, via the local co-ordinates ξ_m and ξ_s, depends on displacements. In Eqs. (5.18) a new notation for the linearisation of the local co-ordinates, given earlier by (2.27), was introduced.

Finally, the variables from the voltage gap g_V with respect to voltages can be determined as

$$\begin{aligned} \Delta_V g_V &= \Delta_V V_{mn} - \Delta_V V_{sn}, \\ \delta_V g_V &= \delta_V V_{mn} - \delta_V V_{sn}, \\ \Delta_V \delta_V g_V &= 0. \end{aligned} \tag{5.19}$$

The linearisation of the variation of the voltage gap with respect to voltages is zero because in the discretisation of voltage along the beam element a typical approach from the finite element method is used, where the voltage at an arbitrary point of the element is linearly dependent on nodal voltages.

5.3 Weak Formulation of Electro-mechanical Contact

In the analysis of the electro-mechanical contact, besides the principle of virtual work, one must also use an appropriate equation concerning the electric current flow. Hence, the global set of equations for the problem at hand can be written down in the following form

$$\begin{cases} \delta_u \Pi_{mM} + \delta_u \Pi_{sM} + \sum\limits_{\text{act}} \delta_u \Pi_{cM} = 0 \\ \delta_V \Pi_{mE} + \delta_V \Pi_{sE} + \sum\limits_{\text{act}} \delta_V \Pi_{cE} = 0 \end{cases} \tag{5.20}$$

after an appropriate extension of the functional variation (1.10) and taking advantage of the notation of the variations with respect to mechanical and electrical variables. In the relation (5.20) additional subscripts M and E are introduced to distinguish between the components of the functional related to mechanical and electric phenomena occuring in both contacting bodies.

Expressions $\delta_V \Pi_{mE}$ and $\delta_V \Pi_{sE}$ in Eq. $(5.20)_2$ are related to the electric current flow in the conductors m and s, respectively. The resulting matrices of the electric conductivity are derived in Section 5.4.

Components of (5.20) related to the contact in the active points take the form

$$\begin{aligned} \delta_u \Pi_{cM} &= F_N \, \delta_u g_N \, , \\ \delta_V \Pi_{cE} &= I \, \delta_V g_V \, . \end{aligned} \tag{5.21}$$

The linearisation of Eqs. (5.20), necessary to solve this set of non-linear equations by the Newton-Raphson method must be carried out with respect to the mechanical variables – displacements and, separately, to the electric variables – voltages. In this way the following relations are obtained

$$\begin{aligned} \Delta_u \delta_u \Pi_{cM} &= \Delta_u F_N \, \delta_u g_N + F_N \, \Delta_u \delta_u g_N \, , \\ \Delta_V \delta_u \Pi_{cM} &= \Delta_V F_N \, \delta_u g_N + F_N \, \Delta_V \delta_u g_N \, , \\ \Delta_u \delta_V \Pi_{cE} &= \Delta_u I \, \delta_V g_V + I \, \Delta_u \delta_V g_V \, , \\ \Delta_V \delta_V \Pi_{cE} &= \Delta_V I \, \delta_V g_V + I \, \Delta_V \delta_V g_V \, . \end{aligned} \tag{5.22}$$

Equation $(5.22)_1$ concerns the purely mechanical part of the contact formulation, which was presented in Chapter 2. Equations $(5.22)_2$ and $(5.22)_3$ provide components related to the electro-mechanical coupling. However, due to the fact that here the mechanical variables do not depend on voltages, the linearisations with respect to voltages in $(5.22)_2$ vanish. Finally, the last formula $(5.22)_4$ corresponds to the purely electric part of contact.

The component related to the coupling of the fields can be expressed as

$$\Delta_u \delta_V \Pi_{cE} = \frac{\partial I}{\partial g_V} \Delta_u g_V \, \delta_V g_V + \frac{\partial I}{\partial g_N} \Delta_u g_N \, \delta_V g_V + I \, \Delta_u \delta_V g_V \tag{5.23}$$

and taking advantage of the expression of the electric current in the contact (5.15) and its partial derivatives (5.16) one gets

$$\begin{aligned} \Delta_u \delta_V \Pi_{cE} &= 2 K_{eq} \sqrt{\alpha\beta} \, \sqrt[3]{\frac{3 \varepsilon_N r_{eq}}{2 E_{eq}}} \\ &\times \left(\sqrt[3]{g_N} \, \Delta_u g_V \, \delta_V g_V + \frac{g_V}{3 \sqrt[3]{g_N}^{\,2}} \Delta_u g_N \, \delta_V g_V + \sqrt[3]{g_N} \, g_V \, \Delta_u \delta_V g_V \right) . \end{aligned} \tag{5.24}$$

The component corresponding to the purely electric contact takes the form

$$\Delta_V \delta_V \Pi_{cE} = 2K_{eq} \sqrt{\alpha\beta} \sqrt[3]{\frac{3\varepsilon_N g_N r_{eq}}{2E_{eq}}} \Delta_V g_V \; \delta_V g_V \tag{5.25}$$

with the use of the expression $(5.19)_3$.

5.4 Beam Finite Element for the Electric Current Flow

Taking into account of electro-mechanical contact in the finite element analysis results in additional components, which must be included in the stiffness matrices and the electric conductivity matrices for the elements. For the sake of completeness of the considered problem a conductivity matrix for a one-dimensional electric conductor with the direct current is derived in this section.

The fundament for the derivation is the Ohm's law, which expresses the electric current as a product of the electric conductivity h and the voltage gradient V

$$I = h\frac{\partial V}{\partial x}. \tag{5.26}$$

Let us consider an electric conductor with the length L and the constant cross-section area A, with the voltages and currents at the ends V_1, I_1 and V_2, I_2, respectively. The conductor is made of a material with the specific conductivity K. For such a conductor the principle of electric power can be written in the following form

$$I_1\, \delta V_1 + I_2\, \delta V_2 = \int_0^L I\, \delta V\, dx. \tag{5.27}$$

The formula (5.27) is analogous to the principle of virtual temperature used in the problems of heat flow (e.g. Bathe 1996). The terms on the left hand side of (5.27) denote the external virtual electric power. Substituting the expression for the electric current (5.26) and the relation determining the electric conductivity of the analysed conductor

$$h = K\frac{A}{L} \tag{5.28}$$

and taking the constants in front of the integral one gets

$$I_1\, \delta V_1 + I_2\, \delta V_2 = K\frac{A}{L}\int_0^L \frac{\partial V}{\partial x}\, \delta V\, dx. \tag{5.29}$$

Now the approximation of voltage within the element can be carried out. To this end the linear shape functions are used

$$N_1 = \frac{1}{2}(1-\xi),$$
$$N_2 = \frac{1}{2}(1+\xi), \tag{5.30}$$

where the local dimensionless co-ordinate $-1 \le \xi \le 1$ is introduced. Then the voltage in any point of the conductor is expressed as

$$V = N_1 V_1 + N_2 V_2 = \begin{bmatrix} N_1 \\ N_2 \end{bmatrix} \circ \begin{bmatrix} V_1 \\ V_2 \end{bmatrix} = \mathbf{N} \circ \mathbf{V} \,. \tag{5.31}$$

Substitution of this expression to the principle of virtual electric power (5.29) leads to a matrix relation between the electric currents and the voltages at the ends of the conductor in the form

$$\mathbf{I}_M = \begin{bmatrix} I_1 \\ I_2 \end{bmatrix} = \mathbf{h}\,\mathbf{V} = \begin{bmatrix} h_{11} & h_{12} \\ h_{21} & h_{22} \end{bmatrix} \begin{bmatrix} V_1 \\ V_2 \end{bmatrix}. \tag{5.32}$$

The elements of the conductivity matrix $\mathbf{h}$ can be determined from the formula including derivatives of the shape functions (5.30)

$$h_{ij} = K \frac{A}{L} \int_0^L \frac{\partial N_i}{\partial \xi} \frac{\partial N_j}{\partial \xi} \left(\frac{\partial \xi}{\partial x} \right)^2 dx \,. \tag{5.33}$$

In this way the electric conductivity matrix is obtained

$$\mathbf{h} = K \frac{A}{L} \begin{bmatrix} 1 & -1 \\ -1 & 1 \end{bmatrix}, \tag{5.34}$$

in which an analogy to the stiffness matrix of a 1D truss finite element can be found.

5.5 Discretisation and Electro-mechanical Contact Finite Element

Derivation of the tangent stiffness matrix and the residual vector for the electro-mechanical contact presented in this section is general and can be used within the contact formulation without smoothing, discussed in Chapters 2 and 3 or with the smooth contact formulations presented in Chapter 4. Now the vector of the nodal unknowns has the following form

$$\mathbf{q}^T = \left(\mathbf{u}_M{}^T, \mathbf{u}_S{}^T, \mathbf{V}_M{}^T, \mathbf{V}_S{}^T \right), \tag{5.35}$$

where a split into the vectors of the nodal displacements and the nodal voltages for the beams m and s is done. For the case without smoothing and for the node-preserving model of smoothing the displacement vectors for each beam contain 12 entries, (2.29) or (4.28), and for the inscribed curve smoothing – 9 (4.27). The voltage vectors for the case without smoothing and for the node-preserving smoothing have two unknowns, while for the inscribed curve smoothing – three.

Discretisation of the displacement field remains unchanged, the formulae (2.34) to (2.37) are still valid with the matrices **G** and **H** corresponding to a given contact formulation. In discretisation of the voltages one can use the expression (5.31), what leads to the following relations for the variations and the linearisations of voltage

$$\begin{aligned}
\delta V_{mn} &= \mathbf{N}_{mn} \circ \delta \mathbf{V}_M ,\\
\delta V_{sn} &= \mathbf{N}_{sn} \circ \delta \mathbf{V}_S ,\\
\Delta V_{mn} &= \mathbf{N}_{mn} \circ \Delta \mathbf{V}_M ,\\
\Delta V_{sn} &= \mathbf{N}_{sn} \circ \Delta \mathbf{V}_S ,\\
\delta V_{mn,m} &= \mathbf{D} \circ \delta \mathbf{V}_M ,\\
\delta V_{sn,s} &= \mathbf{D} \circ \delta \mathbf{V}_S .
\end{aligned} \tag{5.36}$$

For the case without smoothing the shape functions in the vectors $\mathbf{N}_{mn}$ and $\mathbf{N}_{sn}$ are the same as in (5.30), while for the inscribed curve C^1-continuous contact element, presented in Fig. 4.3, the voltages in an arbitrary point of the smooth curve segment can be determined using the following vectors of shape functions

$$\begin{aligned}
\mathbf{N}_{mn} &= \frac{1}{4}\begin{bmatrix} 1-\xi_{mn} \\ 2 \\ 1+\xi_{mn} \end{bmatrix},\\
\mathbf{N}_{sn} &= \frac{1}{4}\begin{bmatrix} 1-\xi_{sn} \\ 2 \\ 1+\xi_{sn} \end{bmatrix}.
\end{aligned} \tag{5.37}$$

For the case of the node-preserving smooth contact element presented in Fig. 4.4 the appropriate vectors of the shape functions have the same form as for the unsmoothed contact element (5.30).

The vector **D** in $(5.36)_5$ and $(5.36)_6$ groups the derivatives of the shape functions with respect to the local co-ordinates ξ_m or ξ_s. For the adopted linear shape functions these derivatives are constant numbers and no distinction between the vectors **D** for the beams m and s must be introduced.

Discretisation of the electro-mechanical variables, calculated in Section 5.3 is the next step of the derivation. The linearisations (5.18) calculated for the function g_V with respect to displacements take the following form

$$\Delta_u g_V = \left(V_{mn,m}\mathbf{F}_m - V_{sn,s}\mathbf{F}_s\right)\begin{bmatrix}\Delta\mathbf{u}_M\\ \Delta\mathbf{u}_S\end{bmatrix},$$
$$\Delta_u \delta_V g_V = \left(\delta\mathbf{V}_M{}^T, \delta\mathbf{V}_S{}^T\right)\begin{bmatrix}\mathbf{DF}_m\\ -\mathbf{DF}_s\end{bmatrix}\begin{bmatrix}\Delta\mathbf{u}_M\\ \Delta\mathbf{u}_S\end{bmatrix}. \tag{5.38}$$

The vectors $\mathbf{F}_m$ and $\mathbf{F}_s$ present in (5.38) are defined in (2.39) and (A2.3) and have 24 entries for the unsmoothed and the smooth node-preserving element or 18 – for the smooth inscribed curve element.

The linearisation and the variation (5.19) of the function g_V with respect to voltages after discretisation are

$$\Delta_V g_V = \left(\mathbf{N}_{mn}{}^T \quad -\mathbf{N}_{sn}{}^T\right)\begin{bmatrix}\Delta\mathbf{V}_M\\ \Delta\mathbf{V}_S\end{bmatrix},$$
$$\delta_V g_V = \left(\delta\mathbf{V}_M{}^T, \delta\mathbf{V}_S{}^T\right)\begin{bmatrix}\mathbf{N}_{mn}\\ -\mathbf{N}_{sn}\end{bmatrix}. \tag{5.39}$$

Having determined the above presented expressions one can find the final form of the tangent stiffness matrix and the residual vector for the electro-mechanical contact finite element. Discretisation of the linearisation of the weak form (5.22) leads to the tangent stiffness matrix, which can be suitably written down as

$$\mathbf{K}_c = \begin{bmatrix}\mathbf{K}_{cMM} & \mathbf{0}\\ \mathbf{K}_{cEM} & \mathbf{K}_{cEE}\end{bmatrix}. \tag{5.40}$$

In this notation a split into the purely mechanical part represented by the submatrix $\mathbf{K}_{cMM}$, derived in the previous chapters, the purely electric part $\mathbf{K}_{cEE}$, yielding from (5.25) and the three component matrix $\mathbf{K}_{cEM}$

$$\mathbf{K}_{cEM} = \mathbf{K}_{cEM1} + \mathbf{K}_{cEM2} + \mathbf{K}_{cEM3}, \tag{5.41}$$

which represents the unilateral coupling between the electric and mechanical field yielding from (5.24), was introduced.

The purely electric part of the formulation has the form

$$\Delta_V \delta_V \Pi_{cE} = \left(\delta\mathbf{V}_M{}^T, \delta\mathbf{V}_S{}^T\right)\mathbf{K}_{cEE}\begin{bmatrix}\Delta\mathbf{V}_M\\ \Delta\mathbf{V}_S\end{bmatrix} \tag{5.42}$$

and the electric matrix in (5.42) is

$$\mathbf{K}_{cEE} = 2K_{eq}\sqrt{\alpha\beta}\,\sqrt[3]{\frac{3\varepsilon_N g_N r_{eq}}{2E_{eq}}}\begin{bmatrix}\mathbf{N}_{mn}\\ -\mathbf{N}_{sn}\end{bmatrix}\otimes\begin{bmatrix}\mathbf{N}_{mn}\\ -\mathbf{N}_{sn}\end{bmatrix}. \tag{5.43}$$

The part of the formulation related to the unilateral electro-mechanical coupling takes the form

$$\Delta_u \delta_V \Pi_{cE} = \left(\delta\mathbf{V}_M{}^T, \delta\mathbf{V}_S{}^T\right)\mathbf{K}_{cEM}\begin{bmatrix}\Delta\mathbf{u}_M \\ \Delta\mathbf{u}_S\end{bmatrix} \tag{5.44}$$

and the submatrices in (5.41) are

$$\begin{aligned}
\mathbf{K}_{cEM1} &= 2K_{eq}\sqrt{\alpha\beta}\,\sqrt[3]{\frac{3\varepsilon_N g_N r_{eq}}{2E_{eq}}}\begin{bmatrix}\mathbf{N}_{mn} \\ -\mathbf{N}_{sn}\end{bmatrix}\left[V_{mn,m}\mathbf{F}_m - V_{sn,s}\mathbf{F}_s\right],\\
\mathbf{K}_{cEM2} &= \frac{2}{3}K_{eq}\sqrt{\alpha\beta}\,\sqrt[3]{\frac{3\varepsilon_N r_{eq}}{2E_{eq}g_N{}^2}}\begin{bmatrix}\mathbf{N}_{mn} \\ -\mathbf{N}_{sn}\end{bmatrix}\mathbf{n}^T\left[\mathbf{G}_{mn} \quad -\mathbf{G}_{sn}\right],\\
\mathbf{K}_{cEM3} &= 2K_{eq}\sqrt{\alpha\beta}\,\sqrt[3]{\frac{3\varepsilon_N g_N r_{eq}}{2E_{eq}{}^2}}\, g_V\begin{bmatrix}\mathbf{DF}_m \\ -\mathbf{DF}_s\end{bmatrix}.
\end{aligned} \tag{5.45}$$

Discretisation of the mechanical part of the weak formulation $(5.21)_1$ leads to the mechanical residual vector derived in Chapters 2 and 4, which here is denoted as $\mathbf{R}_{cM}$. On the other hand, for the electric part $(5.21)_2$ one gets

$$\delta_V \Pi_{cE} = \left(\delta\mathbf{V}_M{}^T, \delta\mathbf{V}_S{}^T\right)\mathbf{R}_{cE} \tag{5.46}$$

and the electric residual vector takes the form

$$\mathbf{R}_{cE} = 2K_{eq}\sqrt{\alpha\beta}\,\sqrt[3]{\frac{3\varepsilon_N g_N r_{eq}}{2E_{eq}}}\, g_V\begin{bmatrix}\mathbf{N}_{mn} \\ -\mathbf{N}_{sn}\end{bmatrix}. \tag{5.47}$$

Finally, the full residual vector for the electro-mechanical element can be expressed as

$$\mathbf{R}_c = \begin{bmatrix}\mathbf{R}_{cM} \\ \mathbf{R}_{cE}\end{bmatrix}. \tag{5.48}$$

5.6 Numerical Examples

5.6.1 Introduction

In the five presented examples the electro-mechanical contact between beams with circular cross-sections is analyzed. The external influences – displacements and voltages are applied simultaneously in a given number of identical increments. Each beam is discretised with two-node straight beam elements for the linear analysis.

In the presented calculations friction is not included and the data are taken in a way to avoid large relative displacements. Hence, there is no problem with possible discontinuities of the normal vectors between the elements.

The Newton-Raphson method was used in the solution, the energy tolerance of 10^{-20} was adopted. Thanks to the applied consistent linearisation the quadratic convergence was achieved. At each increment no more than 4 iterations were necessary to obtain the equilibrium state.

Numerical data are given without units because the purpose of the calculations is mainly the qualitative analysis of phenomena in the electro-mechanical beam-to-beam contact. However, the full physical sense can be retrieved, if any consistent set of physical units, e.g. SI units, is applied.

The presented graphical representations are restricted to beams axes only. To enhance the clarity of 3D views in the figures, axes of the "hidden" beams are depicted as "broken".

5.6.2 Example 1

In this example electro-mechanical contact between two cantilever beams, which axes in the initial configuration are presented in Fig. 5.2 is analysed. At the ends

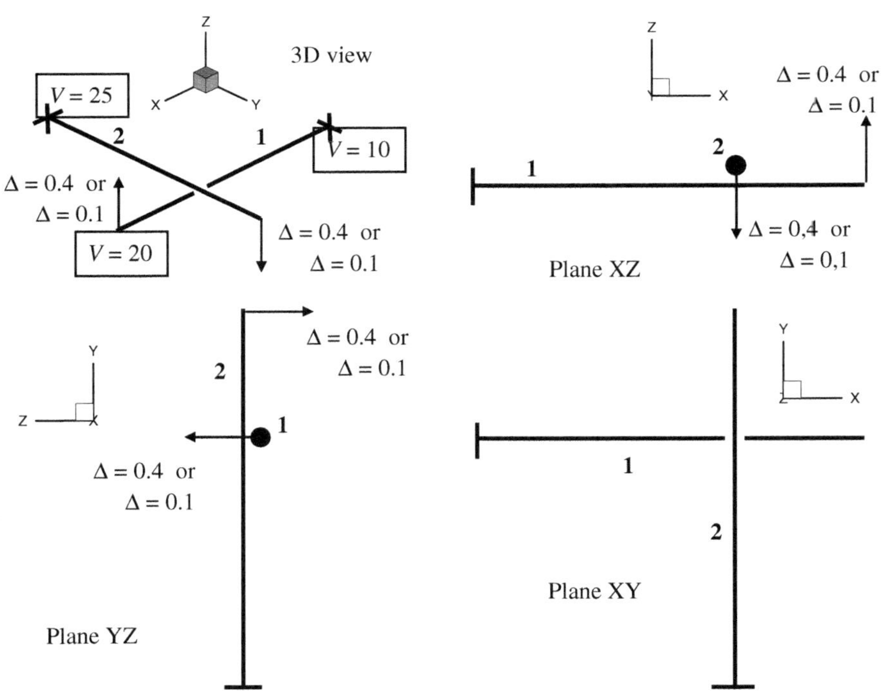

Fig. 5.2 Example 1 – initial configuration of beams axes

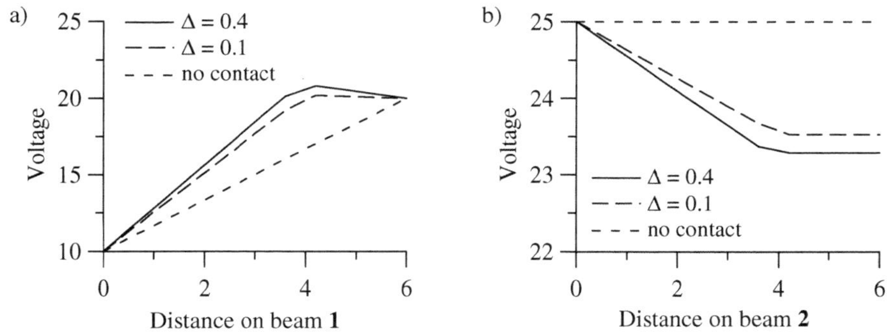

Fig. 5.3 Example 1 – voltage distribution along: a) beam **1**, b) beam**2**

Table 5.1 Example 1 – electric characteristics of contact

	$\Delta = 0.4$	$\Delta = 0.1$
Voltage gap g_V	2.889	3.772
Electric resistance	0.2256	0.3562
Electric current	12.80	10.45

of the beams displacements $\Delta = 0.4$ or 0.1 and voltages – 10 at the support and 20 at the free end of beam **1** and 25 at the support of beam **2**, are applied. Voltages and displacements are applied in 20 increments for the case of displacement $\Delta = 0.4$ or in 10 increments for $\Delta = 0.1$. The following data are used in the calculations: beam **1** – $E = 205{\cdot}10^5$, $\nu = 0.3$, $K = 1.25{\cdot}10^5$, radius of circular cross-section 0.1, length 6,0; beam **2** – $E = 70{\cdot}10^5$, $\nu = 0{,}25$, $K = 4{,}0{\cdot}10^5$, radius of circular cross-section 0.15, length 6,0; initial gap between the beams 0.001.

Graphs showing the distribution of voltage along the conductors for both cases of applied displacements are presented in Fig. 5.3. They are also compared with the linear distribution of voltage, which would be valid for the case without any contact between the beams. The situation of larger imposed displacements is characterised by the larger forces at the contact and larger contact area. Hence, the electric resistance in the contact is smaller, leading to the smaller voltage gap g_V and larger electric current. The smaller value of g_V is manifested by the higher voltage in beam **1** and the smaller voltage in beam **2**. The values of characteristic electric variables in the contact are given in Table 5.1.

5.6.3 Example 2

In this example electro-mechanical contact between two cantilever beams, which axes in the initial configuration are presented in Fig. 5.4 is analysed. At the free end of beam **2** displacement $\Delta = 0.1$ is applied. Besides there are imposed voltages

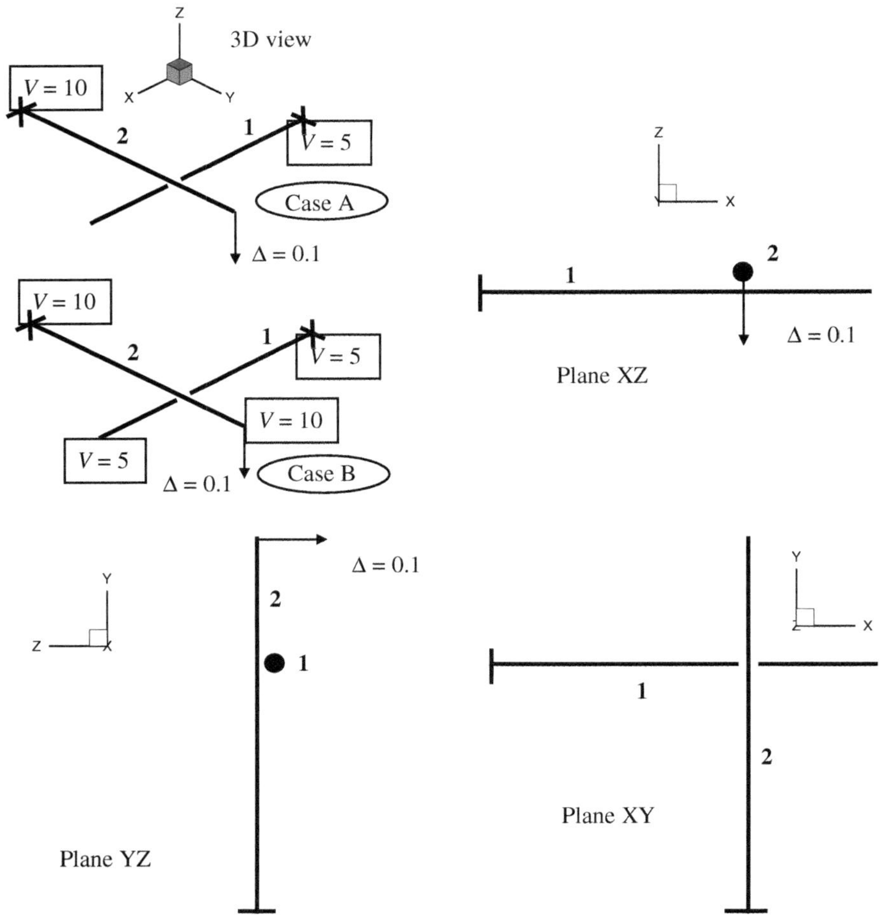

Fig. 5.4 Example 2 – initial configuration of beams axes

at the ends of the beams – 5 for the beam **1** and 10 for beam **2**. In the case A these voltages are applied only at the supports and in the case B – also on the free ends. The displacements and the voltages are applied in 5 increments. The following data, identical for both beams, are used in the calculations – $E = 205{\cdot}10^5$, $\nu = 0.3$, $K = 1.25{\cdot}10^5$, length 6.0, with exception for the radius of the circular cross-section which is 0.1 for beam **1**, and 0.15 for beam **2**. The initial gap between the beams is 0.001.

Graphs showing the distribution of voltage along the conductors for the both cases of the analysed electric boundary conditions are presented in Fig. 5.5. They are also compared with the linear distributions for the case without contact.

In the case B, with the voltages applied at both ends of the beams, a larger voltage gap in the contact than in the case A occurs. Then, with the same value of the electric resistance in the contact, which does not depend on the applied electric boundary conditions, a larger electric current flows through the contact surface. The values of characteristic electric variables in contact are given in Table 5.2.

In this comparison of cases A and B an analogy to the mechanics of statically determinate and statically indeterminate structures can be found. Imposing the electric voltages on both ends of the beams in the case B is an action similar to the addition of extra supports in a statically determinate mechanical system, what leads to redundancy. In the structural engineering such a modified system has a larger stiffness and exhibits smaller displacements. In the presented example "additional" imposed voltages add an electric "stiffness" and lead to smaller differences between the voltage distribution in the situation with and without contact. The result of this is the larger voltage gap in the contact.

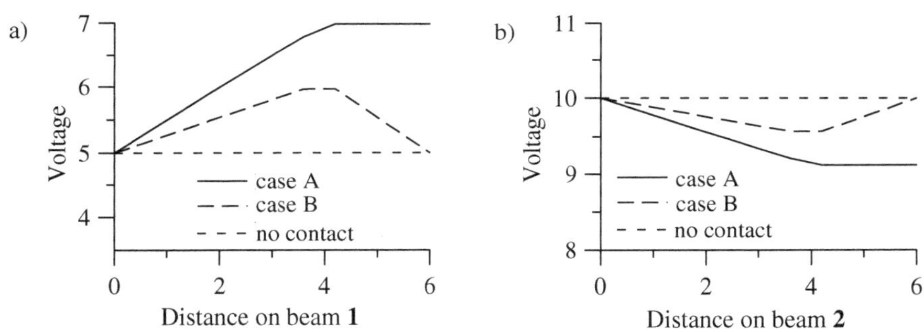

Fig. 5.5 Example 2 – voltage distribution along: a) beam **1**, b) beam **2**

Table 5.2 Example 2 – electric characteristics of contact

	Case A	Case B
Voltage gap g_V	2.224	3.605
Electric resistance	1.141	1.141
Electric current	1.949	3.160

5.6.4 Example 3

In this example electro-mechanical contact between four beams, which axes in the initial configuration are presented in Fig. 5.6 is analysed. The centre points of all beams are partially constrained with the freedom of rotation about the axis lying in the plane XY perpendicular to each beam left. In the case A the imposed displacements $\Delta = 0.4$ are applied at all the beams ends. Besides, at the ends of beam **4** voltages 10 are imposed and at the ends of beam **3** – voltages 20.

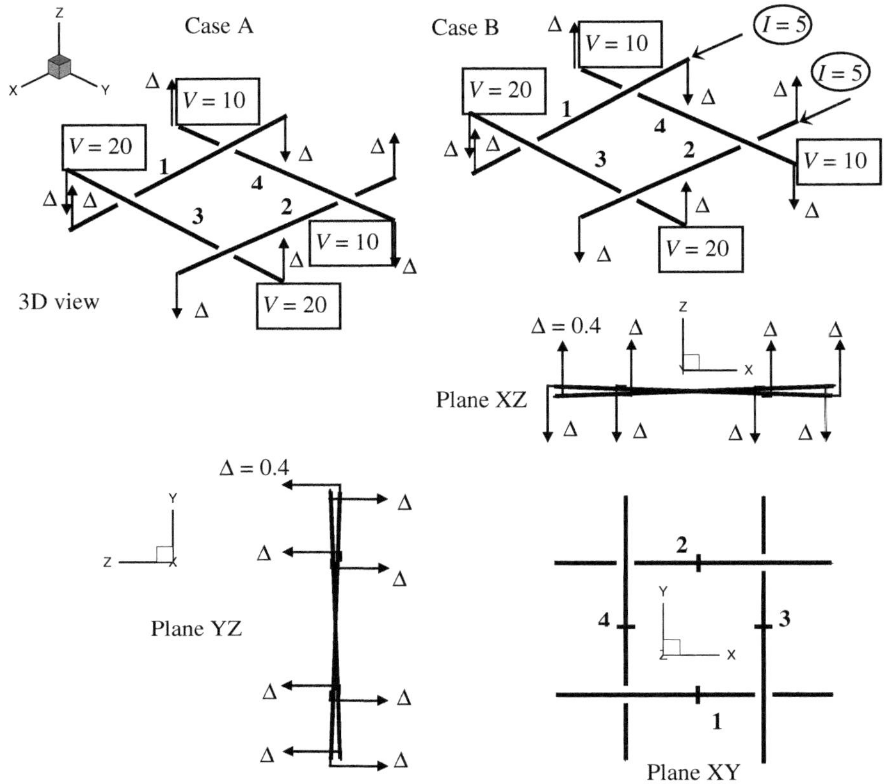

Fig. 5.6 Example 3 – initial configuration of beams axes

Additionally, in the case B at one of the ends of beam **1** and beam **2** electric current 5 is applied. All displacements, voltages and currents are applied in 20 increments. The following data, identical for all the beams are used in the calculations: $E = 205 \cdot 10^5$, $\nu = 0.3$, $K = 1.25 \cdot 10^5$, length 6.0, radius of circular cross-sections 0.1. It is assumed, that in the initial configurations all four contact points are characterised by a penetration 0.05.

Graphs showing the distribution of voltage along conductors **1**, **3** and **4** for the both cases of the analysed electric boundary conditions are presented in Fig. 5.7. A comparison with the linear distribution of voltage for the case without contact is also given.

In this example one can observe the clear influence of the imposed electric currents at the ends of beams **1** and **2**. Their presence results in uneven voltage gaps in the contact points in the case B. On the other hand, in the case A, these gaps are equal in all four contact points and due to this and the symmetry of the system the electric current in all the contacts is the same. In the case B the electric current in

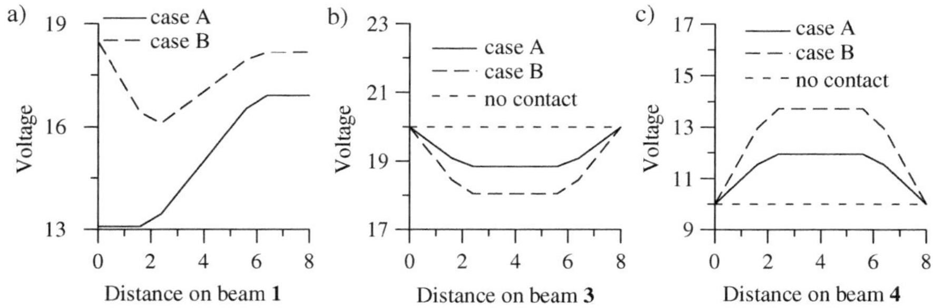

Fig. 5.7 Example 3 – voltage distribution along: a) beam **1**, b) beam **3**, c) beam **4**

Table 5.3 Example 3 – electric characteristics of contact

	Case A	Case B	
		Contact **1–4** and **2–4**	Contact **1–3** and **2–3**
Voltage gap g_V	1.517	2.900	0.901
Electric resistance	0.400	0.400	0.400
Electric current	3.796	7.254	2.254

Table 5.4 Example 3 – convergence analysis

Iteration number	Relative energy	
	10th increment	20th increment
1	1	1
2	7.77×10^{-7}	3.03×10^{-6}
3	9.37×10^{-14}	5.35×10^{-13}
4	1.41×10^{-28}	1.63×10^{-27}

contacts **1–4** and **2–4**, which are located closer to the ends with the imposed electric currents is larger than in the case A, while in contacts **1–3** and **2–3**, located farther from these ends the current is smaller than in the case A. The values of characteristic electric variables in the contact are given in Table 5.3.

It can also be pointed out, that due to the initial penetration between the beams there is no need for separate electric boundary conditions for beams **1** and **2**. If there were no contact in the initial phase, than the resulting set of equations related to the electric analysis would be characterised by a singular matrix of coefficients and no solution would exist. That is why there is no linear voltage distribution for the no-contact situation in the graph in Fig. 5.7a for beam **1**.

Results of convergence analysis for the Newton-Raphson solution method for two selected increments are presented in Table 5.4. Thanks to the consistent linearisation in the formulation of the stiffness and electric conductivity matrices the quadratic convergence is obtained.

5.6.5 Example 4

In this example the influence of angle φ (Fig. 5.1) between the beams axes in the plane XY on the mechanical and electric properties of contact is analysed. Two clamped-clamped beams in a mutually symmetric location, which axes are presented in Fig. 5.8 are considered. At the end of the beams identical displacements $\Delta = 0.2$ and voltages: 10 for beam **1** and 20 for beam **2**, are applied in 20 increments. The following data, identical for all the beams are used in the calculations: $E = 205{\cdot}10^5$, $\nu = 0.3$, $K = 1.25{\cdot}10^5$, length 4.0, radius of circular cross-sections 0.1, initial gap between the beams 0.06.

Mutual relations between the voltage difference g_V, the normal force in contact F_N and the angle φ between the beams axes in the plane XY are presented graphically in Fig. 5.9. The voltage difference depends almost linearly on the normal force. This feature does not change with the change of the angle φ between the beams axes. On the other hand, the relation between the voltage difference and the angle φ is non-linear. The layout of isolines presented in the graph points out that

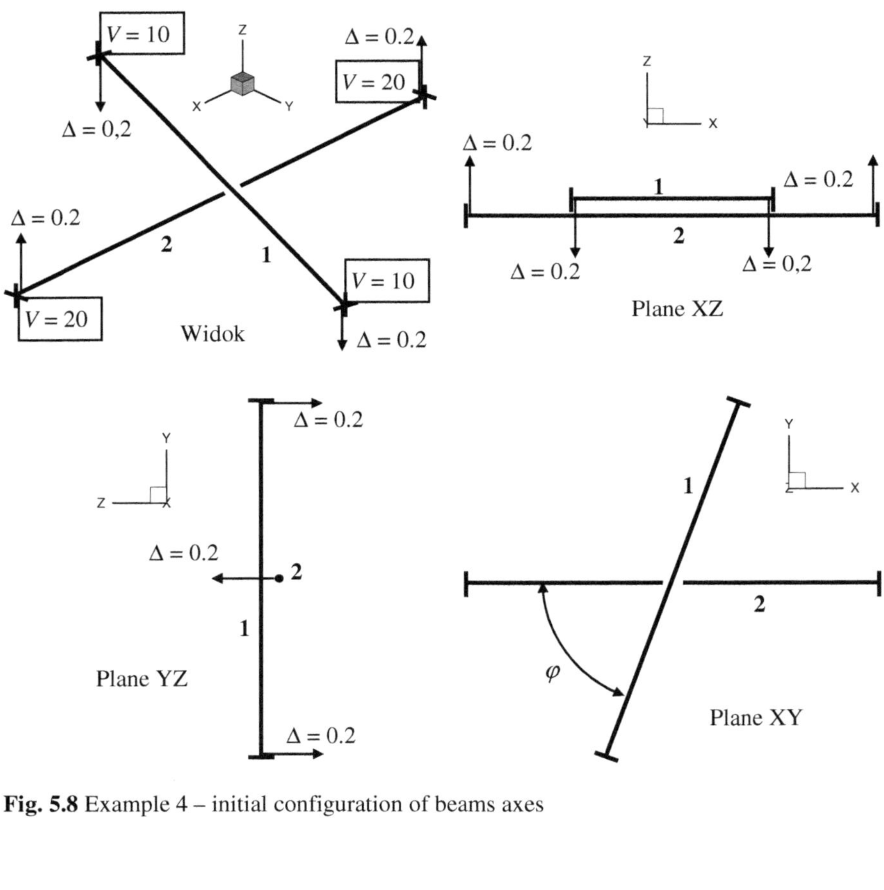

Fig. 5.8 Example 4 – initial configuration of beams axes

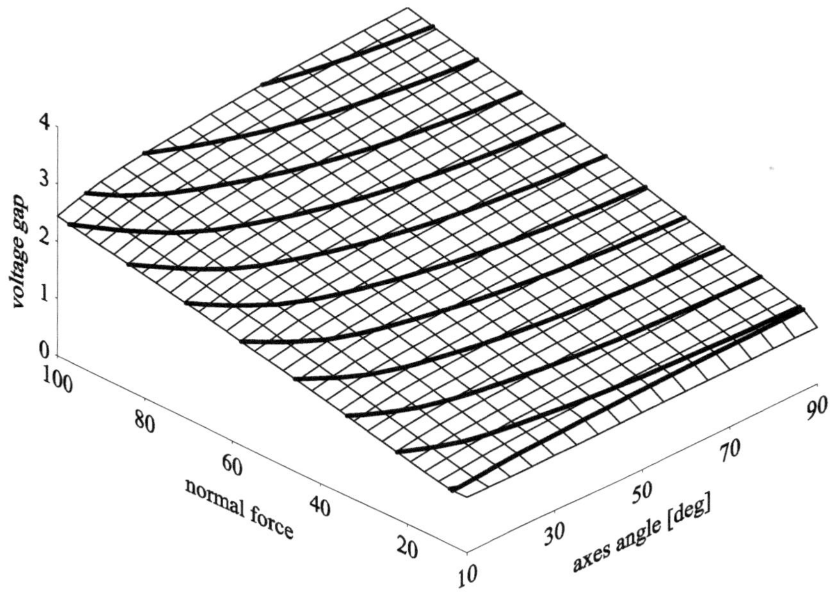

Fig. 5.9 Example 4 – relation between voltage difference, normal force in contact and angle between beams axes in plane

the smaller is the angle the larger is the normal force and also, that the larger is the force the stronger non-linearity in this relation is present. It is especially pronounced for the values of angle φ less than 50°.

5.6.6 Example 5

In this example the electro-mechanical contact between three mutually twisted conductors with axes of a helix shape, presented in Fig. 5.10 is analysed. In the initial configuration there is no contact between the helices. The bottom ends of the conductors are clamped and at the top ends three identical angles of rotation $\phi = 0{,}5$ about axis Z are imposed. They lead to the further twisting of the bundle and bring the conductors into contact. Each of the helices is modelled with 36 straight beam finite elements. As a result the contact, which in reality would occur on a certain length of the helices, is discretised here by contact points between these straight line elements. At the ends of helices voltages are applied. In the case A all three top ends have equal values 20 and the bottom ends – 10. In the case B both ends of helix **1** have voltages 10, helix **2** – 20 and helix **3** – 30. Displacements and voltages are applied in 40 increments. The following data, identical for all the beams are used in the calculations: $E = 205 \cdot 10^5$, $\nu = 0.3$, $K = 1.25 \cdot 10^5$, radius of circular cross-sections 0.1, height of the bundle 10.0, spacing of axes in the plane XY 2.0.

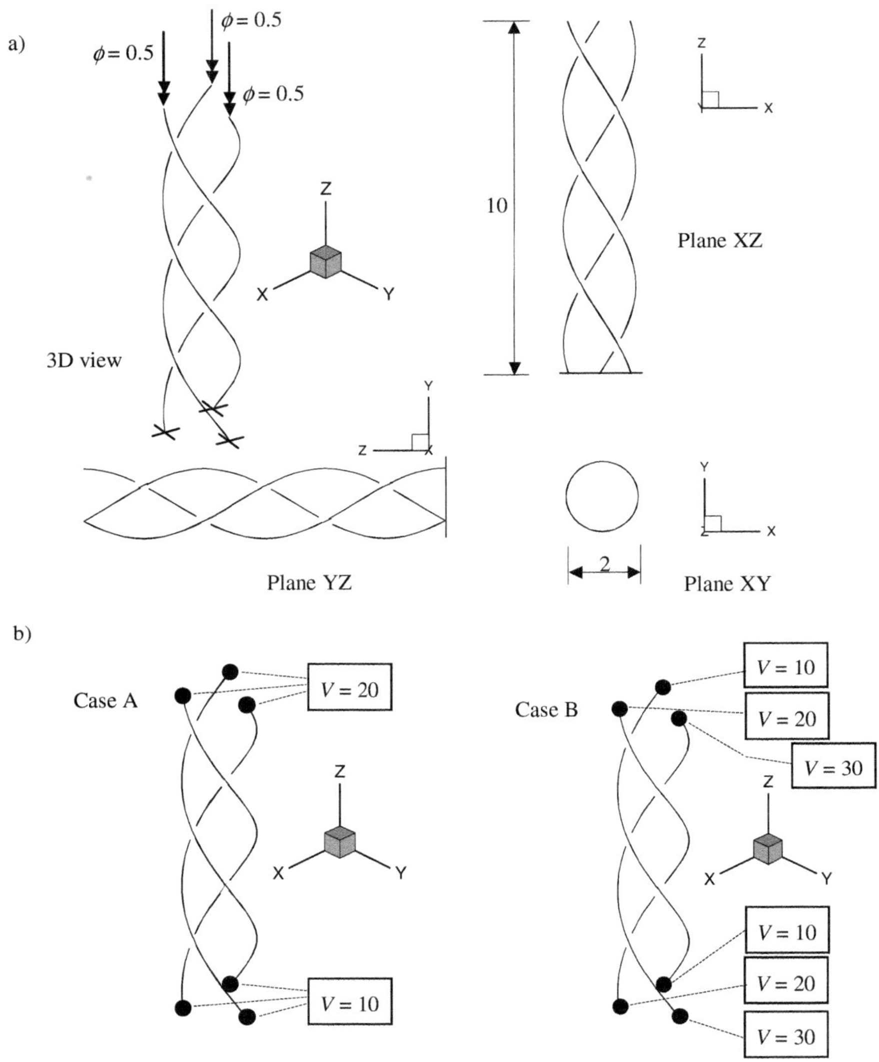

Fig. 5.10 Example 5: a) initial configuration of conductors axes, b) electric boundary conditions

Deformed configuration of the helices axes and graphs showing the voltage distribution along them are presented in Fig. 5.11. It can be noted, that in the case A these distributions only slightly differ from the linear ones valid for the no-contact case. On the other hand, in the case B due to contact a zone is formed, where the voltages of initially different values are equalled and in consequence there is a large discrepancy between the situations with and without contact.

The smooth character of the voltage distributions indicates that the adopted discretization of continuous contact zone by means of contact points can be considered as an attractive alternative for the more exact but very much more expensive analysis, where the conductors would be treated as 3D bodies.

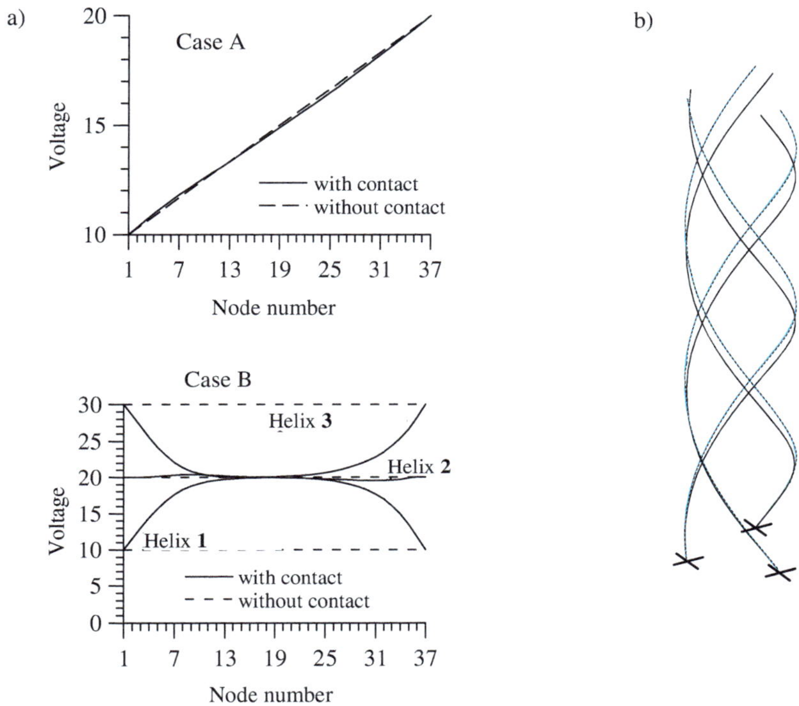

Fig. 5.11 Example 5 – results: a) voltage distributions in the conductors, b) deformed configuration of conductors axes

Chapter 6
Thermo-mechanical Coupling

6.1 Introduction

Analysis of contact with inclusion of coupling between mechanical and thermal fields is a complicated problem because the mutual influences between displacements or strains and temperature are manifested in many different ways. The aspects involved include: heat flow through a real contact area resulting from roughness of contacting surfaces, heat flow through a gas between the bodies, heat flow through radiation; frictional heating; dependence of material properties like elasticity moduli, friction coefficient or heat conduction coefficient on temperature, etc. The more detailed description of various issues related to the thermo-mechanical coupling in contact can be found in the monographs by Wriggers (2002) and Laursen (2002). One can find there numerous references to the papers and other monographs devoted to the problem of the heat conduction in contact. This phenomenon requires a precise description of geometry of bodies surfaces in the micro scale and a development of a thermo-mechanical physical law for the contact. To this end statistical methods can be used (Cooper et al. 1969 and Song and Yovanowic 1987). The numerical solution to this problem was a subject of the papers by Zavarise (1991), Zavarise et al. (1992) as well as Wriggers and Zavarise (1993b). Another problem is related to the frictional heating due to the contact. This topic was considered, for instance, in the papers by Wriggers and Miehe (1992) or by Zavarise et al. (1995, 2005).

Numerical treatment of the thermo-mechanical problem generally depends on the type of the heat flow – steady state, independent of time, or transient one with a variation in time. In the former case the problem is relatively simple and monolithic methods can be effectively used, where both types of unknowns, displacements and temperature, are calculated simultaneously. In the more complicated transient state case, staggered methods are preferred, where the problem is solved iteratively with temperature kept constant and solving for displacements in one iteration and vice versa in the subsequent one. To these methods with respect to the thermo-mechanical contact the papers by Wriggers and Miehe (1992) and by Agelet de Saracibar (1998) were devoted.

In this chapter a formulation of the beam-to-beam contact with a thermo-mechanical coupling in a limited form (Boso et al. 2006) is presented. From the

P. Litewka: Finite Element Analysis of Beam-to-Beam Contact, LNACM 53, pp. 121–134.
springerlink.com

previously mentioned issues of the coupling the heat flow through contact with an assumption of ideally smooth surfaces is taken into account. The heat transfer through the gas and the radiation are neglected. In the physical model of the beams material only the linear thermal expansion is included. All the physical parameters are treated as independent of time.

The analysis presented herein can therefore be considered only as an introduction to a very complicated problem of contact with the coupled fields of displacements and temperature. It can be expanded further, if one takes into account the electric contact discussed in Chapter 5, too. In such a case some additional manifestations of the coupling emerge. They include heat production due to the electric current flow and dependence of electric material and contact properties on temperature.

6.2 Thermo-mechanical Beam Finite Element

Components resulting from contact between bodies are added to the weak formulation independently (2.20) from the ones due to the strain energy, the heat flow or the electric current flow in the contacting bodies. Hence, in the finite element method the contact elements are separate from the finite elements used to the discretisation of bodies themselves and in the beam-to-beam contact the contact element can be attached to any finite element formulation for the beams.

In the analyses presented in this chapter a co-rotational 3D beam finite element derived by Crisfield (1990) is used. However, since one of the aspects considered here is the thermal expansion of the material this feature must be added to the purely mechanical element due to Crisfield. The terms resulting from the steady-state heat flow must be taken into account as well.

It is assumed that the physical parameters of the material do not depend on temperature. Let us denote the constant coefficient of the linear thermal expansion by α_t and by EA – the axial beam stiffness. Then the normal force in a bar subjected to a temperature change can be obtained from the following relation

$$S = EA\,\alpha_t \tau\,, \tag{6.1}$$

where τ is the mean relative temperature in the element related to the initial temperature, for instance to the assembly temperature.

The expanded vector of the degrees of freedom with respect to the Crisfield's element has 14 entries and can be put into the following form

$$\mathbf{q}^T = \left\{\mathbf{u}^T, \boldsymbol{\tau}^T\right\}, \tag{6.2}$$

where the vector of the generalised nodal displacements has the form (2.29) and the vector of the relative nodal temperatures is

$$\boldsymbol{\tau}^T = \left(\tau_1, \tau_2\right). \tag{6.3}$$

With an assumption of linear distribution of temperature along the element and the relation (6.1) the addition to the residual vector in the expanded beam finite element resulting from the heat flow and the thermal expansion can be written down as

$$\mathbf{R}^T = \left\{\mathbf{R}_1^{\ T}, 0, 0, 0, -\mathbf{R}_1^{\ T}, 0, 0, 0, q_{M1}, q_{M2}\right\}. \tag{6.4}$$

The vector $\mathbf{R}_1$ groups the nodal forces resulting from the thermal expansion of the element

$$\mathbf{R}_1^{\ T} = EA\alpha_t\tau\{c_1, c_2, c_3\}. \tag{6.5}$$

The temperature τ is obtained as the mean value of the relative nodal temperatures

$$\tau = \frac{\tau_1 + \tau_2}{2} \tag{6.6}$$

and the values c_i (i = 1, 2 or 3) are the cosines of angles between the current axis of the beam and the axes x_i from the global set of co-ordinates. They can be derived from the relations

$$c_i = \frac{x_{2i} - x_{1i}}{\left\|\mathbf{x}_2 - \mathbf{x}_1\right\|}, \tag{6.7}$$

where $\mathbf{x}_1$ and $\mathbf{x}_2$ are the current position vectors of nodes of the beam finite element. These vectors can be expressed by their components

$$\begin{aligned} \mathbf{x}_1^{\ T} &= \{x_{11}, x_{12}, x_{13}\}, \\ \mathbf{x}_2^{\ T} &= \{x_{21}, x_{22}, x_{23}\}. \end{aligned} \tag{6.8}$$

The quantities denoted with q_{M1} and q_{M2} in the vector (6.4) are the nodal heat fluxes for the considered finite element.

Addition to the tangent stiffness matrix can be put in the form

$$\mathbf{K} = \begin{bmatrix} \mathbf{K}_1 & \mathbf{0} & \mathbf{K}_2 & \mathbf{0} & \mathbf{K}_3 \\ \mathbf{0} & \mathbf{0} & \mathbf{0} & \mathbf{0} & \mathbf{0} \\ -\mathbf{K}_1 & \mathbf{0} & -\mathbf{K}_2 & \mathbf{0} & -\mathbf{K}_3 \\ \mathbf{0} & \mathbf{0} & \mathbf{0} & \mathbf{0} & \mathbf{0} \\ \mathbf{0} & \mathbf{0} & \mathbf{0} & \mathbf{0} & \mathbf{K}_T \end{bmatrix}, \tag{6.9}$$

where the component matrices $\mathbf{K}_1$ and $\mathbf{K}_2$ result from the dependence of the cosines (6.7) on the position vectors (6.8) and, hence, on the nodal displacements. They have the form

$$\mathbf{K}_1 = \begin{bmatrix} \frac{\partial c_1}{\partial u_{11}} & \frac{\partial c_1}{\partial u_{12}} & \frac{\partial c_1}{\partial u_{13}} \\ \frac{\partial c_2}{\partial u_{11}} & \frac{\partial c_2}{\partial u_{12}} & \frac{\partial c_2}{\partial u_{13}} \\ \frac{\partial c_3}{\partial u_{11}} & \frac{\partial c_3}{\partial u_{12}} & \frac{\partial c_3}{\partial u_{13}} \end{bmatrix}, \quad \mathbf{K}_2 = \begin{bmatrix} \frac{\partial c_1}{\partial u_{21}} & \frac{\partial c_1}{\partial u_{22}} & \frac{\partial c_1}{\partial u_{23}} \\ \frac{\partial c_2}{\partial u_{21}} & \frac{\partial c_2}{\partial u_{22}} & \frac{\partial c_2}{\partial u_{23}} \\ \frac{\partial c_3}{\partial u_{21}} & \frac{\partial c_3}{\partial u_{22}} & \frac{\partial c_3}{\partial u_{23}} \end{bmatrix}. \tag{6.10}$$

Here the notation u_{ij}, for the components of the nodal displacements is introduced. These displacements correspond to the nodal position vectors (6.8). The partial derivatives in the matrices (6.10) can be obtained using the symbolic algebra program Maple 7.

The matrix $\mathbf{K}_3$ present in Eq. (6.9) groups the terms resulting from the dependence of the normal force (6.1) on the nodal temperatures and can be expressed as

$$\mathbf{K}_3 = \begin{bmatrix} \frac{c_1}{2} & \frac{c_1}{2} \\ \frac{c_2}{2} & \frac{c_2}{2} \\ \frac{c_3}{2} & \frac{c_3}{2} \end{bmatrix}. \tag{6.11}$$

The matrix $\mathbf{K}_T$ from (6.9) requires a separate comment. In order to derive its components one can use the analogy to the matrix of electric conductivity $\mathbf{h}$ (5.34), presented in Section 5.4, where the Ohm's law (5.26) and the principle of virtual electric powers (5.27) were used. The similar procedure can be used with the equation of the one-dimensional steady-state heat flow

$$q = -k \frac{\partial \tau}{\partial x}, \tag{6.12}$$

where k denotes the heat conduction coefficient for the material, which here is assumed as constant, independent of time. The counterpart of the principle of virtual electric powers is the principle of virtual temperature (Bathe 1996), according to which one can write down the following relation for a finite element

$$q_1\, \delta\tau_1 + q_2\, \delta\tau_2 = k \int_0^L \frac{\partial \tau}{\partial x}\, \delta\tau\, dx\,. \tag{6.13}$$

With this in hand and with the assumption of linear distribution of temperature along the element the matrix of thermal conduction can be found in following form

$$\mathbf{K}_T = k\begin{bmatrix} 1 & -1 \\ -1 & 1 \end{bmatrix}. \tag{6.14}$$

This matrix, like the electric conductivity matrix (5.34), is analogous to the stiffness matrix for the 1D truss finite element.

6.3 Variables for Thermo-mechanical Contact

Similarly as in the case of electro-mechanical contact described in Chapter 5, also in the case of thermo-mechanical contact an additional variable characterizing the thermal field at the contact point must be introduced. It is the temperature, denoted by $\tau(\xi)$, which is assumed to be constant in the cross-section. Then the temperature difference at the contact between two beams can be determined using the temperature values τ_{mn} and τ_{sn} for the cross-sections corresponding to the contact points C_{mn} and C_{ms}

$$g_H = \tau_{mn} - \tau_{sn}. \tag{6.15}$$

The heat flow in contact spots is a complicated phenomenon. It depends on the micro-scale structure of the surfaces, the normal force and the temperature gradient. There are many proposals to formulate physical laws for this phenomenon (Wriggers 2002, Zavarise et al. 1992, Wriggers and Zavarise 1993b). In the presented analyses a very simple model is assumed, in which only an unlimited heat flow through the contact between ideally smooth surfaces is considered. In such a case, besides the constraints resulting from the purely mechanical aspects of contact, discussed in Chapters 1, 2 and 3, the thermal constraint is also imposed in the following form

$$g_H = 0. \tag{6.16}$$

Such a constraint enforces the equal temperatures for the both cross-sections of the beams at the contact point. This constraint is enforced using the penalty method, which yields the following expression for the heat flow

$$q = \varepsilon_H g_H, \tag{6.17}$$

where the thermal penalty parameter ε_H is introduced.

It is worth to note that, Eq. (6.17) can also be seen as a thermo-mechanical physical law for the heat flow in the contact, if instead of a constant and possibly large value of the penalty parameter ε_H one assumes a physical value, which could be derived from experimental analyses. Such a physical parameter may also be variable, depending on temperature, normal force in contact or degree of surfaces roughness. However, in the presented approach this value is taken as constant and an analysis of its influence on numerical results is presented in one of the examples solved in Section 6.6.

In the weak form and its linearisation the variation, the linearisation and the linearisation of the variation of the temperature difference (6.15) calculated with respect to mechanical unknowns (displacements) and thermal unknowns (temperatures) are necessary. In order to distinguish between them, in the notation for these operators the subscripts, u for the displacements and T for the temperatures, are introduced.

The variables related to the displacements can be expressed as

$$\begin{aligned} \Delta_u g_H &= \Delta_u(\tau_{mn} - \tau_{sn}) = \tau_{mn,m}\, \Delta_u \xi_{mn} - \tau_{sn,s}\, \Delta_u \xi_{sn} , \\ \Delta_u \delta_T g_H &= \delta_T(\Delta_u g_H) = \delta_T \tau_{mn,m}\, \Delta_u \xi_{mn} - \delta_T \tau_{sn,s}\, \Delta_u \xi_{sn} . \end{aligned} \tag{6.18}$$

In this derivation the dependence of temperature in contact on the local coordinate, which in turn is a function of displacements, was taken into account.

The variables related to the temperature take the form

$$\begin{aligned} \delta_T g_H &= \delta_T \tau_{mn} - \delta_T \tau_{sn} , \\ \Delta_T g_H &= \Delta_T \tau_{mn} - \Delta_T \tau_{sn} , \\ \Delta_T \delta_T g_H &= 0 . \end{aligned} \tag{6.19}$$

The linearisation of the variation of the temperature difference vanishes because in the adopted finite element model the approximation of temperature along the element is a linear function of the nodal temperatures.

The general form of the thermo-mechanical contact formulation allows to introduce the mutual relations between displacement and temperature fields. However, within the assumptions given in Section 6.1, the penetration function g_N, discussed in Section 2.2 depends only on displacements, not on temperature. That is why the variables calculated from this gap function with respect to temperature are zero

$$\begin{aligned} \Delta_T g_N &= 0 , \\ \Delta_T \delta_u g_N &= 0 . \end{aligned} \tag{6.20}$$

6.4 Weak Form for Thermo-mechanical Contact

In order to carry out the analysis of a system with the thermo-mechanical coupling one must consider, besides the principle of virtual work, also an appropriate equation for the heat flow. Hence, after an expansion of the variation of the functional (1.10) and introduction of the separate notation for the variations with respect to displacements and temperatures, one can get the global set of equations for the thermo-mechanical contact in the following form

$$\begin{cases} \delta_u \Pi_{mM} + \delta_u \Pi_{sM} + \sum\limits_{\text{act}} \delta_u \Pi_{cM} = 0 \\ \delta_T \Pi_{mH} + \delta_T \Pi_{sH} + \sum\limits_{\text{act}} \delta_T \Pi_{cH} = 0 \end{cases} \tag{6.21}$$

The additional subscripts M and H are introduced in (6.21) to distinguish between the components of the functional resulting from the mechanical and thermal phenomena occurring in both beams.

The components $\delta_T\Pi_{mH}$ and $\delta_T\Pi_{sH}$ in Eq. $(6.21)_2$ are related to the heat flow through the beams m and s. The resulting matrices of the heat conductivity are described in Section 6.2. The component of $(6.21)_2$, resulting from the contact in the active points takes the form

$$\delta_T \Pi_{cH} = q\, \delta_T g_H \,. \tag{6.22}$$

The linearisation of Eqs. (6.21), necessary for the Newton-Raphson solution method of the non-linear set of equations of contact with thermo-mechanical coupling must be carried out separately for both groups of unknowns: mechanical – displacements and thermal – temperatures. In this way the following relations are obtained

$$\begin{aligned} \Delta_u \delta_u \Pi_{cM} &= \Delta_u F_N\, \delta_u g_N + F_N\, \Delta_u \delta_u g_N \,, \\ \Delta_T \delta_u \Pi_{cM} &= \Delta_T F_N\, \delta_u g_N + F_N\, \Delta_T \delta_u g_N \,, \end{aligned} \qquad (6.23)_{1,2}$$

$$\begin{aligned} \Delta_u \delta_T \Pi_{cH} &= \Delta_u q\, \delta_T g_H + q\, \Delta_u \delta_T g_H \,, \\ \Delta_T \delta_T \Pi_{cH} &= \Delta_T q\, \delta_T g_H + q\, \Delta_T \delta_T g_H \,. \end{aligned} \qquad (6.23)_{3,4}$$

The formula $(6.23)_1$ gives the purely mechanical part of the contact formulation, which was derived in Section 2. The relations $(6.23)_2$ and $(6.23)_3$ lead to the elements responsible for the thermo-mechanical coupling. However, due to the adopted assumption, that the mechanical properties do not depend on temperature, the linearisations in $(6.23)_2$ vanish. Finally, the last of the relations $(6.23)_4$ gives the purely thermal part of the contact.

If the advantage is taken of the formula for the heat transfer in contact in the form (6.17), then the component responsible for the coupling of the mechanical and thermal fields has the form

$$\Delta_u \delta_T \Pi_{cH} = \varepsilon_H \Delta_u g_H\, \delta_T g_H + \varepsilon_H g_H\, \Delta_u \delta_V g_V \,. \tag{6.24}$$

The component responsible for the purely thermal part of contact, taking into account Eq. $(6.19)_3$, can be written down as

$$\Delta_T \delta_T \Pi_{cH} = \varepsilon_H\, \Delta_T g_H\, \delta_T g_H \,. \tag{6.25}$$

6.5 Discretisation and Thermo-mechanical Contact Beam Finite Element

Similarly as in the case of the electro-mechanical contact finite element presented in Section 5.5, derivation of the tangent stiffness matrix and the residual vector for the thermo-mechanical beam-to-beam contact finite element has a general character and can be used within the contact formulation without smoothing presented in Chapter 2 and 3 or for the C^1-continuous contact elements derived in Chapter 4.

The vector of nodal unknowns can be set in the following form

$$\mathbf{q}^T = \left(\mathbf{u}_M{}^T, \mathbf{u}_S{}^T, \boldsymbol{\tau}_M{}^T, \boldsymbol{\tau}_S{}^T\right), \tag{6.26}$$

where a split into the nodal displacement and the nodal temperature vectors for the beams m and s was carried out. For the case without smoothing and for the node-preserving smooth elements the displacement vector for each beam has 12 entries, (2.29) or (4.28) and for the inscribed curve smooth element – 9 (4.27). The temperature vectors have respectively 2 or 3 elements.

For the discretisation of the displacement field the formulae (2.34) to (2.37) are valid with the matrices **G** and **H**, appropriate for the contact formulation. The discretisation of the temperature field can be assumed in the same way as the voltages are discretised in Section 5.5 (5.31)

$$\tau = N_1\tau_1 + N_2\tau_2 = \begin{bmatrix} N_1 \\ N_2 \end{bmatrix} \circ \begin{bmatrix} \tau_1 \\ \tau_2 \end{bmatrix} = \mathbf{N} \circ \boldsymbol{\tau}\,, \tag{6.27}$$

what leads to the following expressions for the discretised variations and linearisations of temperature

$$\begin{aligned}
\delta\tau_{mn} &= \mathbf{N}_{mn} \circ \delta\boldsymbol{\tau}_M\,,\\
\delta\tau_{sn} &= \mathbf{N}_{sn} \circ \delta\boldsymbol{\tau}_S\,,\\
\Delta\tau_{mn} &= \mathbf{N}_{mn} \circ \Delta\boldsymbol{\tau}_M\,,\\
\Delta\tau_{sn} &= \mathbf{N}_{sn} \circ \Delta\boldsymbol{\tau}_S\,,\\
\delta\tau_{mn,m} &= \mathbf{D} \circ \delta\boldsymbol{\tau}_M\,,\\
\delta\tau_{sn,s} &= \mathbf{D} \circ \delta\boldsymbol{\tau}_S\,.
\end{aligned} \tag{6.28}$$

In the case without smoothing and for the node-preserving smooth element the shape functions in the vectors $\mathbf{N}_{mn}$ and $\mathbf{N}_{sn}$ are as in (5.30) and for the inscribed curve smooth element the temperature in an arbitrary point on the smooth curve can be calculated using Eq. (5.37). The vector **D** groups the derivatives of the shape functions with respect to the local co-ordinates ξ_m or ξ_s.

Now, the discretised thermo-mechanical variables discussed in Section 6.3 can be calculated. The linearisations (6.18) derived for the function g_H with respect to displacements can be expressed as

$$\begin{aligned}
\Delta_u g_H &= \left(\tau_{mn,m}\mathbf{F}_m - \tau_{sn,s}\mathbf{F}_s\right)\begin{bmatrix} \Delta\mathbf{u}_M \\ \Delta\mathbf{u}_S \end{bmatrix},\\
\Delta_u \delta_T g_H &= \left(\delta\boldsymbol{\tau}_M{}^T, \delta\boldsymbol{\tau}_S{}^T\right)\begin{bmatrix} \mathbf{DF}_m \\ -\mathbf{DF}_s \end{bmatrix}\begin{bmatrix} \Delta\mathbf{u}_M \\ \Delta\mathbf{u}_S \end{bmatrix},
\end{aligned} \tag{6.29}$$

where the vectors $\mathbf{F}_m$ and $\mathbf{F}_s$ are determined in the same way as for Eqs. (5.38). The discretised linearisation and the variation (6.19) for g_H with respect to temperature read

$$\Delta_T g_H = \left(\mathbf{N}_{mn}{}^T \quad -\mathbf{N}_{sn}{}^T\right)\begin{bmatrix}\Delta\boldsymbol{\tau}_M \\ \Delta\boldsymbol{\tau}_S\end{bmatrix}, \tag{6.30$_1$}$$

$$\delta_T g_H = \left(\delta\boldsymbol{\tau}_M{}^T, \delta\boldsymbol{\tau}_S{}^T\right)\begin{bmatrix}\mathbf{N}_{mn} \\ -\mathbf{N}_{sn}\end{bmatrix}. \tag{6.30$_2$}$$

The last stage in the development of the thermo-mechanical contact element is derivation of the tangent stiffness matrix and the residual vector. Discretisation of the linearisation of the weak form (6.23) leads to the matrix, which can be expressed in the following form

$$\mathbf{K}_c = \begin{bmatrix}\mathbf{K}_{cMM} & \mathbf{0} \\ \mathbf{K}_{cHM} & \mathbf{K}_{cHH}\end{bmatrix}, \tag{6.31}$$

where a split into the purely mechanical part represented by the matrix $\mathbf{K}_{cMM}$, derived in previous chapters, the purely thermal part given in the matrix $\mathbf{K}_{cHH}$, resulting from the Eq (6.25), and the two-component matrix $\mathbf{K}_{cHM}$

$$\mathbf{K}_{cHM} = \mathbf{K}_{cHM1} + \mathbf{K}_{cHM2} \tag{6.32}$$

representing the unilateral thermo-mechanical coupling resulting from the discretisation of Eq. (6.24) is introduced.

The thermal part of the formulation after discretisation takes the form

$$\Delta_T \delta_T \Pi_{cH} = \left(\delta\boldsymbol{\tau}_M{}^T, \delta\boldsymbol{\tau}_S{}^T\right)\mathbf{K}_{cHH}\begin{bmatrix}\Delta\boldsymbol{\tau}_M \\ \Delta\boldsymbol{\tau}_S\end{bmatrix} \tag{6.33}$$

and the thermal contact matrix is

$$\mathbf{K}_{cHH} = \varepsilon_H \begin{bmatrix}\mathbf{N}_{mn} \\ -\mathbf{N}_{sn}\end{bmatrix} \otimes \begin{bmatrix}\mathbf{N}_{mn} \\ -\mathbf{N}_{sn}\end{bmatrix}. \tag{6.34}$$

The part of the formulation due to the unilateral thermo-mechanical coupling is given by

$$\Delta_u \delta_T \Pi_{cH} = \left(\delta\boldsymbol{\tau}_M{}^T, \delta\boldsymbol{\tau}_S{}^T\right)\mathbf{K}_{cHM}\begin{bmatrix}\Delta\mathbf{u}_M \\ \Delta\mathbf{u}_S\end{bmatrix} \tag{6.35}$$

and the component matrices in (6.32) are

$$\mathbf{K}_{cHM1} = \varepsilon_H \begin{bmatrix} \mathbf{N}_{mn} \\ -\mathbf{N}_{sn} \end{bmatrix} \left[\tau_{mn,m} \mathbf{F}_m - \tau_{sn,s} \mathbf{F}_s \right], \tag{6.36$_1$}$$

$$\mathbf{K}_{cHM2} = \varepsilon_H g_H \begin{bmatrix} \mathbf{DF}_m \\ -\mathbf{DF}_s \end{bmatrix}. \tag{6.36$_2$}$$

Discretisation of the thermal part of the weak form (6.22) can be expressed as

$$\delta_T \Pi_{cH} = \left(\delta\boldsymbol{\tau}_M{}^T, \delta\boldsymbol{\tau}_S{}^T \right) \mathbf{R}_{cH} \tag{6.37}$$

and the included thermal residual vector takes the form

$$\mathbf{R}_{cH} = \varepsilon_H g_H \begin{bmatrix} \mathbf{N}_{mn} \\ -\mathbf{N}_{sn} \end{bmatrix}. \tag{6.38}$$

Finally, the entire residual vector for the thermo-mechanical contact element can be set up as

$$\mathbf{R}_c = \begin{bmatrix} \mathbf{R}_{cM} \\ \mathbf{R}_{cH} \end{bmatrix}, \tag{6.39}$$

where $\mathbf{R}_{cM}$ denotes the mechanical part of the residual vector, which was derived in Chapters 2 or 4.

As was previously mentioned, the presented thermo-mechanical contact formulation can be joined with the electro-mechanical contact element presented in Chapter 5. In such a case the formulation would consist of three equations. i.e. both equations (5.20) and Eq. (6.21)$_2$. Then the linearisation would be carried out separately for three groups of unknowns: displacements, temperatures and voltages. Discretisation would lead to the tangent stiffness matrix for the contact element with the three-field coupling in the form

$$\mathbf{K}_c = \begin{bmatrix} \mathbf{K}_{cMM} & \mathbf{K}_{cME} & \mathbf{K}_{cMH} \\ \mathbf{K}_{cEM} & \mathbf{K}_{cEE} & \mathbf{K}_{cEH} \\ \mathbf{K}_{cHM} & \mathbf{K}_{cHE} & \mathbf{K}_{cHH} \end{bmatrix}, \tag{6.40}$$

which would correspond to the following vector of element unknowns

$$\mathbf{q}^T = \left(\mathbf{u}_M{}^T, \mathbf{u}_S{}^T, \mathbf{V}_M{}^T, \mathbf{V}_S{}^T, \boldsymbol{\tau}_M{}^T, \boldsymbol{\tau}_S{}^T \right). \tag{6.41}$$

The component matrices in (6.40) result from various aspects of the thermo-electro-mechanical coupling. For instance, the heat production due to the electric current flow is included in $\mathbf{K}_{cHE}$, dependence of the electric conductivity on temperature – in $\mathbf{K}_{cEH}$, frictional heating – as additional terms in $\mathbf{K}_{cHM}$, and dependence of mechanical properties of contact on temperature – in $\mathbf{K}_{cMH}$. The inclusion of these issues is beyond the scope of the present considerations and are left as an attractive topic for the further research.

6.6 Numerical Examples

6.6.1 Introduction

In this section two numerical examples are presented to illustrate the properties of the simplified thermo-mechanical beam-to-beam contact element derived above. Frictionless contact of beams with circular cross-sections is analyzed. Each beam is discretised with 10 finite elements presented in Section 6.2.

Analyzed beams are subjected to imposed displacements and temperature boundary conditions. These external influences are applied simultaneously in a given number of equal increments.

The energy tolerance in the iterative Newton-Raphson method is equal to 10^{-20}.

Similarly as in all examples presented in previous chapters the numerical data are given without any units. In order to add the full physical meaning to these analyses any consistent set of units can be incorporated, e.g. the basic SI units.

The presented graphical representations are restricted to beams axes only. To enhance the 3D views in the figures axes of "hidden" beams are depicted as "broken".

6.6.2 Example 1

In this example contact between two cantilever beams, which axes in the initial configuration are shown in Fig. 6.1, is considered. At the free end of beam **1** displacement $\Delta = 0.5$ is applied. Besides, there are imposed temperatures on the beams ends: – 0 and 5 on beam **1** as well as 50 and 45 on beam **2**. Displacements and temperature are applied in 20 increments. The following data are used in the calculations: beam **1** – $E = 200{\cdot}10^5$, $\nu = 0.3$, $k = 1$, $\alpha_t = 1.2{\cdot}10^{-5}$, radius of circular cross-section 0.1, length 6.0; beam **2** – $E = 75{\cdot}10^5$, $\nu = 0.2$, $k = 1$, $\alpha_t = 1.2{\cdot}10^{-5}$, radius of circular cross-section 0.1, length 6.0; initial gap between the beams 0.01; normal contact penalty parameter $\varepsilon_N = 1000$. In this example an influence of the thermal penalty parameter on the results is investigated. Three different values are taken: $\varepsilon_H = 1$, 10 or 100.

Deformed configuration of the beams axes and graphs of the temperature distribution along the beams are presented in Fig. 6.2. In these graphs the influence of the thermal penalty parameter, which plays a role of the heat conduction coefficient in the contact, can be observed. Its increase leads to equalling of temperature in the beams cross-sections at the contact. The temperature difference decreases from 12.51 for $\varepsilon_H = 1$, through 1.72 for $\varepsilon_H = 10$ to 0.18 for $\varepsilon_H = 100$. The linear

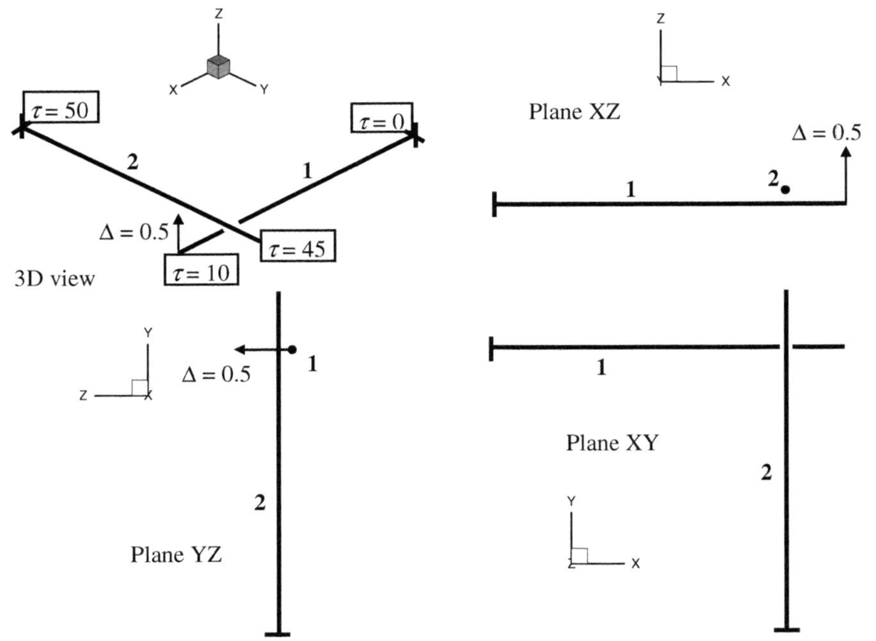

Fig. 6.1 Example 1 – initial configuration of beams axes

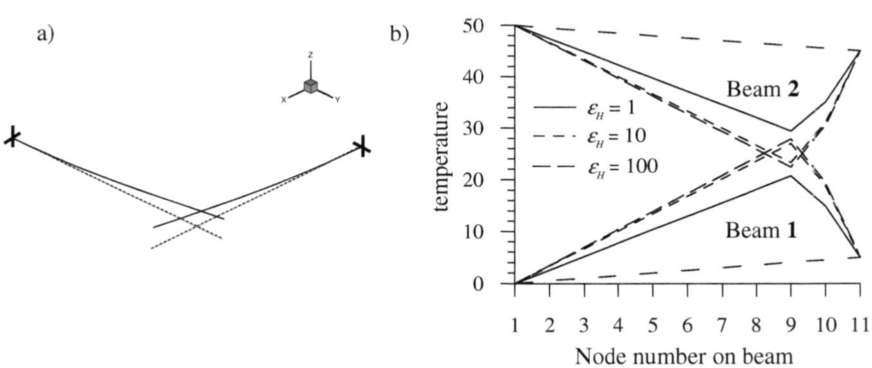

Fig. 6.2 Example 1: a) deformed configuration of beams axes, b) temperature distribution

temperature distribution for the no-contact case is also presented in the graphs. The discrepancy between it and the presented non-linear distributions for the contact case is an effect of the displacements influence and the resulting contact between the beams on the temperature field. The opposite influence does not exist directly in the considered contact formulation due to the adopted assumptions. Still, this influence can be observed in Fig. 6.3, where the displacements of points

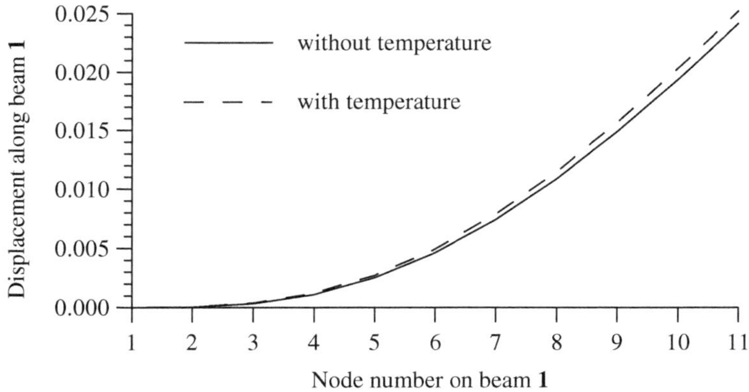

Fig. 6.3 Example 1 – temperature influence on beam **1** displacements

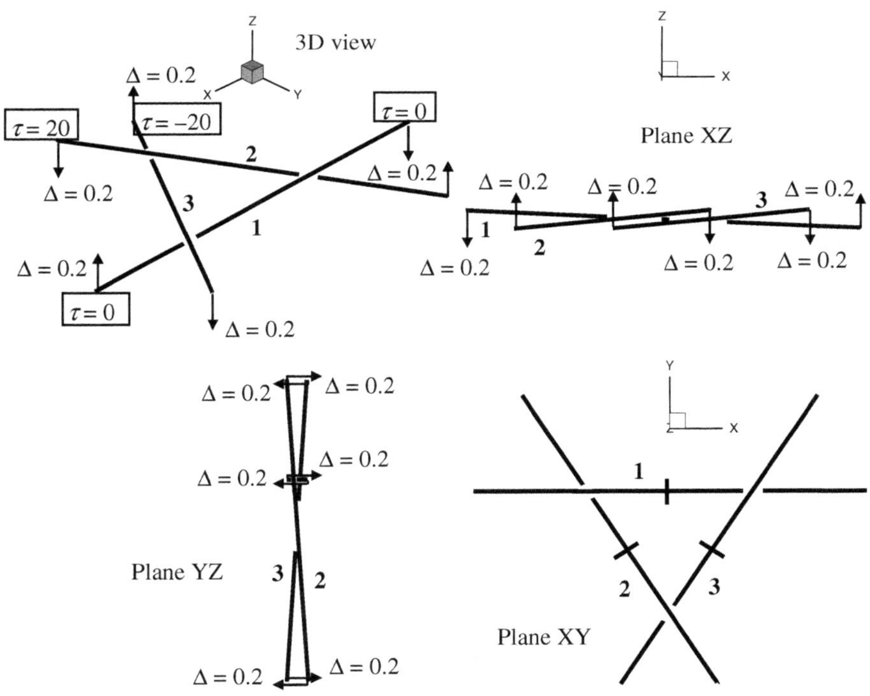

Fig. 6.4 Example 2 – initial configuration of beams axes

on beam **1** along its initial axis (X) for the case $\varepsilon_H = 100$ with and without temperature are presented. It can be seen that the temperature field, modified by the contact feeds back to the system and modifies displacements. In this sense the thermo-mechanical coupling in the considered form is bilateral.

6.6.3 Example 2

In this example contact between three beams, which axes in the initial configuration are presented in Fig. 6.4 is analysed. The centre points of all three beams are fully clamped. At the free ends of the beams displacements $\Delta = 0.2$ leading to mutual contact are applied. Besides, there are imposed temperatures: 0 at the both ends of beam **1**, 20 at one end of beam **2** and –20 at one end of beam **3**. Displacements and temperatures are applied in 20 increments. The following data, identical for all the beams, are taken in the calculations: $E = 200 \cdot 10^5$, $\nu = 0.3$, $k = 1$, $\alpha_t = 1.2 \cdot 10^{-5}$, radius of circular cross-section 0.1, length 6.0. The penalty parameters are: $\varepsilon_N = 10000$, $\varepsilon_H = 100$. In the initial configuration there is the penetration between the beams: for the pairs **1–3** and **2–3** – 0.0509, and for the pair **1–2** – 0.000996.

Results of calculations – the deformed configuration and the temperature distribution along the beams, are presented in Fig. 6.5. The temperature difference in the contact between the beams is: for the pair **1–2** – 0.0093, and for the pairs **1–3** and **2–3** – 0.0398.

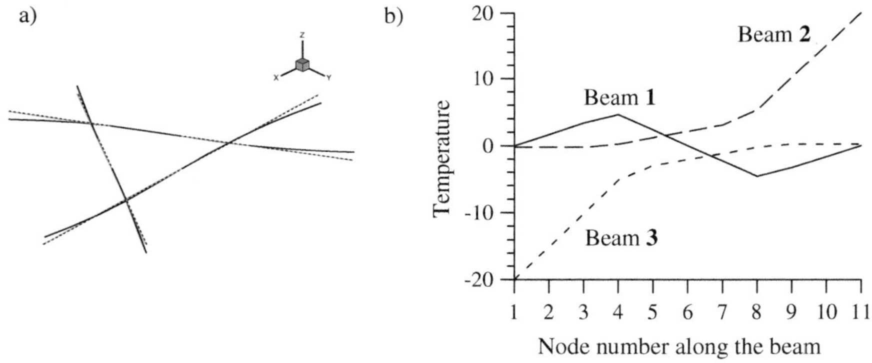

Fig. 6.5 Example 2: a) deformed configuration of beams axes, b) temperature distribution

Chapter 7
Summary and Outlook

Beam-to-beam contact is a special case among the broad range of possibilities of contact between bodies, that are continuously discussed and analysed in the engineering literature. It requires specific procedures of contact search, unique criteria of contact existence and definition of the contact finite element. The scope of literature on this topic is relatively small. The aim of the present work was to fill this gap.

In the presented analyses results of multi-aspect contact between beams are presented. The main purpose was to derive consistent contact finite elements, basing on the full linearisation of the weak form, for all the considered cases of contact. The elements were tested and their efficiency was checked in a series of purposefully chosen numerical examples.

In the first part, in Chapters 2 and 3 the formulation of contact with friction for beams with rectangular cross-sections was presented. The contact constraints were introduced using the penalty and Lagrange multipliers methods. An original contact search routine was presented together with a contact existence criterion for beams with rectangular cross-sections. Two methods of sliding check for the Coulombs law were presented – separately for two friction forces acting along the contacting beams and one check for the case, when these forces are treated as components of one resultant friction force. Effectiveness of the formulation was confirmed in several numerical examples including multiple beams contact. A comparison for both presented methods used to include the contact constraints was carried out and their known advantages and disadvantages were confirmed. It was strongly pointed out, that the penalty method is much easier to use. Another comparison with a full 3D analysis carried out by the program Abaqus was also done and very satisfactory results were obtained.

Another part of the work in Chapter 4 was devoted to the formulation of C^1-continuous contact finite elements. Contrary to the first part, where the beam finite elements and contact formulation used the same shape functions, here the contact was treated completely separately from the analysis of beams themselves. All four types of elements presented can be attached to any formulation of beams modelled with any type of beam elements. Two types of curve construction were considered – inscribed curve and node-preserving. Within each of these constructions Hermite's polynomials and Bezier's curves were used to approximate the curve segments. Respective contact finite elements for the consistent, full linearisation were derived using the symbolic algebra program Maple 7. In several numerical examples, featuring large sliding between the adjacent curve segments, the effectiveness of all four elements was confirmed as far as the smoothing is concerned. However,

P. Litewka: Finite Element Analysis of Beam-to-Beam Contact, LNACM 53, pp. 135–136.
springerlink.com

it was found, that the Hermite inscribed curve element, despite quite complicated and long code for its characteristic matrices, exhibited the best performance with the fastest convergence and, consequently, the shortest computer times.

In the last part of the work, in Chapters 5 and 6, some aspects of thermo-electro-mechanical coupling in contact were analysed. From the numerous mutual relations between the considered fields only few were taken into account, i.e.: relation between the electric resistivity of contact and the normal contact force, influence of the contact point location on the temperature and voltage distribution and thermal expansion of beams. Within these assumptions contact finite elements were derived and applied in a couple of numerical examples. Among others, a mutual relation between the normal contact force, voltage gap and angle between beams axes was determined. It was pointed out that in the assumed range of assumptions the electro-mechanical contact features a unilateral coupling and the thermo-mechanical contact – bilateral coupling.

The full analysis of the extremely complex problem of thermo-electro-mechanical contact requires inclusion of many other factors. Hence, the considerations presented in this work can be treated only as an introduction to the main purpose of the multi-aspect contact analysis, for instance for a case of electric current flow through a bundle of multi-level twisted conductors or superconductors. To this end the future research should be devoted to the following issues:

1. Introduction of variable electric and thermal fields and addition of alternating current flow.
2. Taking into account the heating due to the electric current flow. This would require more sophisticated analysis methods, since the consistent linearisation of this transient problem and application of monolithic method seems too complex to be carried out.
3. Inclusion of dependence of material parameters on temperature. It especially concerns the friction coefficient, the heat conduction coefficient and the moduli of elasticity. It is also advisable to use a more sophisticated model of friction.
4. Derivation and introduction of more sophisticated thermo-mechanical law for the beam-to-beam contact. However, it is understood that a too complex microscopic law would lead to much higher complexity of the formulation and all virtues of the beam formulation as far as its simplicity is concerned would be lost in comparison to the full 3D analysis. It is still purposeful to find this physical law in a form similar to the one used in the electro-mechanical contact.
5. Taking into account the coupling with magnetic fields and resulting additional forces acting on the system.

The above mentioned problems indicate that the beam-to-beam contact formulation can be developed further to gain the full insight into the phenomena occurring in the case of complex processes in the modern engineering, where the influence zones of the classical branches of physics like: mechanics, electricity, thermomechanics and magnetism overlap and call for collaboration between various specialists.

Appendix 1
Matrices D and E for Beams with Rectangular Cross-Sections

A1.1 Components of Matrix D

In the following derivations subscripts m and s are omitted to simplify the notation.

The matrix **D**, present in Eqs. (2.30) and (2.31), can be expressed in the following form

$$\mathbf{D} = \left[d_{jk}\right]_{3\times 12} \tag{A1.1}$$

and its components in each row j are given by the formulae

$$\begin{aligned}
d_{1k} &= N_1\big|_{k=1} + N_2\big|_{k=7} + A_{1,k}\sin\varphi + \\
&\quad + A_1\varphi_{,k}\cos\varphi + A_{2,k}(1-\cos\varphi) + A_2\varphi_{,k}\sin\varphi, \\
d_{2k} &= N_3^s\big|_{k=2} + N_5^s\big|_{k=6} + N_4^s\big|_{k=8} + N_6^s\big|_{k=12} + \\
&\quad + A_{3,k}\sin\varphi + A_3\varphi_{,k}\cos\varphi + A_{4,k}(1-\cos\varphi) + A_4\varphi_{,k}\sin\varphi, \\
d_{3k} &= N_3^t\big|_{k=3} + N_5^t\big|_{k=5} + N_4^t\big|_{k=9} + N_6^t\big|_{k=11} + \\
&\quad + A_{5,k}\sin\varphi + A_5\varphi_{,k}\cos\varphi + A_{6,k}(1-\cos\varphi) + A_6\varphi_{,k}\sin\varphi.
\end{aligned} \tag{A1.2}$$

In Eqs. (A1.2) the notation

$$(\ )_{,k} = \frac{\partial(\ )}{\partial u_k}$$

for partial derivatives was used and the symbol $(\ )\big|_c$ means that the component in the parenthesis is taken into account, only when the condition c is fulfilled. To simplify the notation further, the bars over the displacements u_k and angles φ present in Eq. (2.16), denoting that the corresponding values are measured in the local co-ordinate system, are also omitted.

The further notation used in Eqs. (A1.2) is

$$A_1 = \frac{-s\varphi_t + t\varphi_s}{\varphi},$$

$$A_2 = \frac{\varphi_r(s\varphi_s + t\varphi_t)}{\varphi^2},$$

$$A_3 = \frac{-t\varphi_r}{\varphi}.$$

Derivatives of the angle φ calculated with respect to the local co-ordinates u_k can be expressed as

$$\varphi_{,1} = \varphi_{,7} = 0, \quad \varphi_{,4} = \frac{\varphi_r}{\varphi} N_1, \quad \varphi_{,10} = \frac{\varphi_r}{\varphi} N_2,$$

$$\varphi_{,2} = -\varphi_{,8} = \frac{-\varphi_t}{\varphi} N_7^s, \quad \varphi_{,6} = \frac{\varphi_t}{\varphi} N_8^s, \quad \varphi_{,12} = \frac{\varphi_t}{\varphi} N_9^s,$$

$$\varphi_{,3} = -\varphi_{,9} = \frac{\varphi_s}{\varphi} N_7^t, \quad \varphi_{,5} = \frac{\varphi_s}{\varphi} N_8^t, \quad \varphi_{,11} = \frac{\varphi_s}{\varphi} N_9^t;$$

and of the coefficients A_1 to A_6 –

$$A_{1,1} = A_{1,7} = 0, \quad A_{1,4} = \frac{q_1\varphi_r N_1}{\varphi}, \quad A_{1,10} = \frac{q_1\varphi_r N_2}{\varphi},$$

$$A_{1,2} = -A_{1,8} = \left(\frac{s}{\varphi} - q_1\varphi_t\right)N_7^s, \quad A_{1,6} = -\left(\frac{s}{\varphi} - q_1\varphi_t\right)N_8^s, \quad A_{1,12} = -\left(\frac{s}{\varphi} - q_1\varphi_t\right)N_9^s$$

$$A_{1,3} = -A_{1,9} = \left(\frac{t}{\varphi} + q_1\varphi_s\right)N_7^t, \quad A_{1,5} = \left(\frac{t}{\varphi} + q_1\varphi_s\right)N_8^t, \quad A_{1,11} = \left(\frac{t}{\varphi} + q_1\varphi_s\right)N_9^t,$$

$$A_{2,1} = A_{2,7} = 0, \quad A_{2,4} = \left(\varphi^2 - 2\varphi_r^{\,2}\right)q_2 N_1, \quad A_{2,3} = -A_{2,9} = -\left(2q_2\varphi_r\varphi_s - \frac{s\varphi_r}{\varphi^2}\right)N_7^t,$$

$$A_{2,2} = -A_{2,8} = \left(2q_2\varphi_r\varphi_t - \frac{t\varphi_r}{\varphi^2}\right)N_7^s, \quad A_{2,6} = -\left(2q_2\varphi_r\varphi_t - \frac{t\varphi_r}{\varphi^2}\right)N_8^s,$$

$$A_{2,12} = -\left(2q_2\varphi_r\varphi_t - \frac{t\varphi_r}{\varphi^2}\right)N_9^s, \quad A_{2,10} = \left(\varphi^2 - 2\varphi_r^{\,2}\right)q_2 N_2,$$

$$A_{2,5} = -\left(2q_2\varphi_r\varphi_s - \frac{s\varphi_r}{\varphi^2}\right)N_8^t, \quad A_{2,11} = -\left(2q_2\varphi_r\varphi_s - \frac{s\varphi_r}{\varphi^2}\right)N_9^t$$

$$A_{3,1} = A_{3,7} = 0, \quad A_{3,4} = \left(\varphi_r q_3 - \frac{1}{\varphi}\right)tN_1, \quad A_{3,10} = \left(\varphi_r q_3 - \frac{1}{\varphi}\right)tN_2,$$

$$A_{3,2} = -A_{3,8} = t\varphi_t q_3 N_7^s, \quad A_{3,6} = t\varphi_t q_3 N_8^s, \quad A_{3,12} = t\varphi_t q_3 N_9^s,$$

$$A_{3,3} = -A_{3,9} = t\varphi_s q_3 N_7^t, \quad A_{3,11} = t\varphi_s q_3 N_9^t, \quad A_{3,5} = t\varphi_s q_3 N_8^t;$$

$$A_{4,1} = A_{4,7} = 0, \quad A_{4,4} = -(2sq_3\varphi + \varphi_r q_{42})N_1, \quad A_{4,10} = -(2sq_3\varphi + \varphi_r q_{42})N_2,$$

$$A_{4,2} = -A_{4,8} = (\varphi_t q_{42} - q_{41})N_7^s, \quad A_{4,6} = -(\varphi_t q_{42} - q_{41})N_8^s,$$

$$A_{4,12} = -(\varphi_t q_{42} - q_{41})N_9^s, \quad A_{4,3} = -A_{4,9} = \left(\frac{t\varphi_t}{\varphi^2} - \varphi_s q_{42}\right)N_7^t,$$

$$A_{4,5} = \left(\frac{t\varphi_t}{\varphi^2} - \varphi_s q_{42}\right)N_8^t, \quad A_{4,11} = \left(\frac{t\varphi_t}{\varphi^2} - \varphi_s q_{42}\right)N_9^t;$$

$$A_{5,k} = \frac{-s}{t}A_{3,k};$$

$$A_{6,1} = A_{6,7} = 0, \quad A_{6,4} = -(2tq_3\varphi + \varphi_r q_{62})N_1, \quad A_{6,10} = -(2tq_3\varphi + \varphi_r q_{62})N_2,$$

$$A_{6,2} = -A_{6,8} = \left(\varphi_t q_{62} - \frac{s\varphi_s}{\varphi^2}\right)N_7^s, \quad A_{6,6} = \left(\frac{s\varphi_s}{\varphi^2} - \varphi_t q_{62}\right)N_8^s,$$

$$A_{6,12} = \left(\frac{s\varphi_s}{\varphi^2} - \varphi_t q_{62}\right)N_9^s, \quad A_{6,3} = -A_{6,9} = (q_{61} - \varphi_s q_{62})N_7^t$$

$$A_{6,5} = (q_{61} - \varphi_s q_{62})N_8^t, \quad A_{6,11} = (q_{61} - \varphi_s q_{62})N_9^t,$$

where the following notation was introduced

$$q_1 = \frac{s\varphi_t - t\varphi_s}{\varphi^3}, \quad q_2 = \frac{s\varphi_s + t\varphi_t}{\varphi^4}, \quad q_3 = \frac{\varphi_r}{\varphi^3},$$

$$q_{41} = \frac{t\varphi_s - 2s\varphi_t}{\varphi^2}, \quad q_{42} = \frac{2\left[t\varphi_s\varphi_t - s\left(\varphi_t^2 + \varphi_r^2\right)\right]}{\varphi^4},$$

$$q_{61} = \frac{s\varphi_t - 2t\varphi_s}{\varphi^2}, \qquad q_{62} = \frac{2\left[s\varphi_s\varphi_t - t\left(\varphi_s{}^2 + \varphi_r{}^2\right)\right]}{\varphi^4}.$$

A1.2 Components of Matrix E

The matrix **E** present in Eqs. (2.32) and (2.33) can be expressed in the form

$$\mathbf{E} = \left[e_{jk}\right]_{3\times 12} \tag{A1.3}$$

and its components in each row j are given by the formulae

$$\begin{aligned}
e_{1k} &= N_1\big|_{k=1} + N_2\big|_{k=7} + \left(A_{1,k\xi} - A_1\varphi_{,k}\varphi_{,\xi}\right)\sin\varphi + \\
&+ \left(A_{1,k}\varphi_{,\xi} + A_{1,\xi}\varphi_{,k} + A_1\varphi_{,k\xi}\right)\cos\varphi + \left(A_{2,k}\varphi_{,\xi} + A_{2,\xi}\varphi_{,k} + A_2\varphi_{,k\xi}\right)\sin\varphi + \\
&+ A_{2,k\xi}(1-\cos\varphi) + A_2\varphi_{,k}\varphi_{,\xi}\cos\varphi, \\
e_{2k} &= N_3^s\big|_{k=2} + N_5^s\big|_{k=6} + N_4^s\big|_{k=8} + N_6^s\big|_{k=12} + \\
&+ \left(A_{4,k}\varphi_{,\xi} + A_{4,\xi}\varphi_{,k} + A_4\varphi_{,k\xi}\right)\sin\varphi + A_{4,k\xi}(1-\cos\varphi) + \\
&+ A_4\varphi_{,k}\varphi_{,\xi}\cos\varphi + \left(A_{3,k}\varphi_{,\xi} + A_{3,\xi}\varphi_{,k} + A_3\varphi_{,k\xi}\right)\cos\varphi + \\
&+ \left(A_{3,k\xi} - A_3\varphi_{,k}\varphi_{,\xi}\right)\sin\varphi, \\
e_{3k} &= N_3^t\big|_{k=3} + N_5^t\big|_{k=5} + N_4^t\big|_{k=9} + N_6^t\big|_{k=11} + \\
&+ \left(A_{6,k}\varphi_{,\xi} + A_{6,\xi}\varphi_{,k} + A_6\varphi_{,k\xi}\right)\sin\varphi + A_{6,k\xi}(1-\cos\varphi) + \\
&+ A_6\varphi_{,k}\varphi_{,\xi}\cos\varphi + \left(A_{5,k}\varphi_{,\xi} + A_{5,\xi}\varphi_{,k} + A_5\varphi_{,k\xi}\right)\cos\varphi + \\
&+ \left(A_{5,k\xi} - A_5\varphi_{,k}\varphi_{,\xi}\right)\sin\varphi.
\end{aligned} \tag{A1.4}$$

Eqs. (A1.4) contain derivatives with respect to the local co-ordinate ξ of the quantities defined in Section A1.1. The explicit form of these expressions is very long and they are not presented here. They can be calculated using basic, though tedious operations.

Appendix 2
Derivation of Variables $\Delta\delta\xi_{mn}$ and $\Delta\delta\xi_{sn}$

The starting point in the derivation of these kinematic variables is the set of orthogonality conditions (2.2). In the first step their variation is calculated, what yields the following set of equations

$$\begin{cases} \delta\mathbf{x}_{mn,m} \cdot \mathbf{x}_{ms} + \mathbf{x}_{mn,m} \cdot \left(\delta\mathbf{x}_{mn} - \delta\mathbf{x}_{sn}\right) = 0 \\ \delta\mathbf{x}_{sn,s} \cdot \mathbf{x}_{ms} + \mathbf{x}_{sn,s} \cdot \left(\delta\mathbf{x}_{mn} - \delta\mathbf{x}_{sn}\right) = 0 \end{cases}, \tag{A2.1}$$

where

$$\mathbf{x}_{ms} = \mathbf{x}_{mn} - \mathbf{x}_{sn}\,.$$

Besides, the variations of the position vector derivatives can be found using the following relations

$$\begin{aligned} \delta\mathbf{x}_{mn,m} &= \mathbf{x}_{mn,mm} \cdot \delta\xi_{mn} + \delta\mathbf{u}_{mn,m}\,, \\ \delta\mathbf{x}_{sn,s} &= \mathbf{x}_{sn,ss} \cdot \delta\xi_{sn} + \delta\mathbf{u}_{sn,s}\,. \end{aligned} \tag{A2.2}$$

After the finite element discretisation the solution of the set of equations (A2.1) takes the form (2.40). In order to simplify the notation it can also be presented as follows

$$\begin{bmatrix} \delta\xi_{mn} \\ \delta\xi_{sn} \end{bmatrix} = \begin{bmatrix} \mathbf{F}_m \\ \mathbf{F}_s \end{bmatrix} \begin{bmatrix} \delta\mathbf{u}_M \\ \delta\mathbf{u}_S \end{bmatrix}. \tag{A2.3}$$

The inverse of the matrix **A** (2.27), after its split into components can be written down as

$$\mathbf{A}^{-1} = \begin{bmatrix} a_{mm} & a_{ms} \\ a_{sm} & a_{ss} \end{bmatrix}. \tag{A2.4}$$

The next step of the derivation is calculation of the linearisations of the set of equations (A2.1). This yields two equations

$$\begin{aligned}
&\Delta\delta\mathbf{x}_{mn,m}\cdot\mathbf{x}_{ms}+\delta\mathbf{x}_{mn,m}\cdot(\Delta\mathbf{x}_{mn}-\Delta\mathbf{x}_{sn})+\Delta\mathbf{x}_{mn,m}\cdot(\delta\mathbf{x}_{mn}-\delta\mathbf{x}_{sn})+\\
&\qquad+\mathbf{x}_{mn,m}\cdot(\Delta\delta\mathbf{x}_{mn}-\Delta\delta\mathbf{x}_{sn})=0,\\
&\Delta\delta\mathbf{x}_{sn,s}\cdot\mathbf{x}_{ms}+\delta\mathbf{x}_{sn,s}\cdot(\Delta\mathbf{x}_{mn}-\Delta\mathbf{x}_{sn})+\Delta\mathbf{x}_{sn,s}\cdot(\delta\mathbf{x}_{mn}-\delta\mathbf{x}_{sn})+\\
&\qquad+\mathbf{x}_{sn,s}\cdot(\Delta\delta\mathbf{x}_{mn}-\Delta\delta\mathbf{x}_{sn})=0.
\end{aligned}\tag{A2.5}$$

The undefined expressions for the derivatives with respect to the local coordinates of the variation linearisations of the position vectors, present in the Eqs. (A2.5), read

$$\begin{aligned}
\Delta\delta\mathbf{x}_{mn,m}&=\mathbf{x}_{mn,mmm}\delta\xi_{mn}\Delta\xi_{mn}+\Delta\mathbf{u}_{mn,mm}\delta\xi_{mn}+\delta\mathbf{u}_{mn,mm}\Delta\xi_{mn}+\\
&\quad+\mathbf{x}_{mn,mm}\Delta\delta\xi_{mn}+\Delta\delta\mathbf{u}_{mn,m},\\
\Delta\delta\mathbf{x}_{sn,s}&=\mathbf{x}_{sn,sss}\delta\xi_{sn}\Delta\xi_{sn}+\Delta\mathbf{u}_{sn,ss}\delta\xi_{sn}+\delta\mathbf{u}_{sn,ss}\Delta\xi_{sn}+\\
&\quad+\mathbf{x}_{sn,ss}\Delta\delta\xi_{sn}+\Delta\delta\mathbf{u}_{sn,s}.
\end{aligned}\tag{A2.6}$$

Now the formulae (2.24), (A2.2), its counterparts for the linearisation and (A2.6) can be substituted to Eqs. (A2.5). This yields a set of equations, which can be given in the following matrix form

$$\mathbf{A}\begin{bmatrix}\Delta\delta\xi_{mn}\\ \Delta\delta\xi_{sn}\end{bmatrix}+\begin{bmatrix}R_m\\ R_s\end{bmatrix}=\begin{bmatrix}0\\ 0\end{bmatrix}.\tag{A2.7}$$

With the simplifying notation

$$\begin{aligned}
p_1&=\mathbf{x}_{sn,sss}\circ\mathbf{x}_{ms}-3\mathbf{x}_{sn,ss}\circ\mathbf{x}_{sn,s},\\
p_2&=\mathbf{x}_{sn,ss}\circ\mathbf{x}_{mn,m},\\
p_3&=\mathbf{x}_{sn,s}\circ\mathbf{x}_{mn,mm},\\
p_4&=\mathbf{x}_{mn,mmm}\circ\mathbf{x}_{ms}+3\mathbf{x}_{mn,mm}\circ\mathbf{x}_{mn,m}
\end{aligned}$$

the components R_m and R_s present in (A2.7) can be expressed as

$$\begin{aligned}
R_m&=p_2\,\delta\xi_{sn}\,\Delta\xi_{sn}+p_3(\delta\xi_{mn}\,\Delta\xi_{sn}+\delta\xi_{sn}\,\Delta\xi_{mn})+p_4\,\delta\xi_{mn}\,\Delta\xi_{mn}+\\
&+\mathbf{x}_{ms}\circ\Delta\delta\mathbf{u}_{mn,m}+\delta\mathbf{u}_{mn,m}\circ(\Delta\mathbf{u}_{mn}-\Delta\mathbf{u}_{sn})+\\
&+(\delta\mathbf{u}_{mn}-\delta\mathbf{u}_{sn})\circ\Delta\mathbf{u}_{mn,m}+\mathbf{x}_{mn,m}\circ(\Delta\delta\mathbf{u}_{mn}-\Delta\delta\mathbf{u}_{sn})+\\
&+\delta\xi_{mn}\left[\mathbf{x}_{ms}\circ\Delta\mathbf{u}_{mn,mm}+\mathbf{x}_{mn,mm}\circ(\Delta\mathbf{u}_{mn}-\Delta\mathbf{u}_{sn})+2\mathbf{x}_{mn,m}\circ\Delta\mathbf{u}_{mn,m}\right]\\
&+\left[\mathbf{x}_{ms}\circ\delta\mathbf{u}_{mn,mm}+\mathbf{x}_{mn,mm}\circ(\delta\mathbf{u}_{mn}-\delta\mathbf{u}_{sn})+2\mathbf{x}_{mn,m}\circ\delta\mathbf{u}_{mn,m}\right]\Delta\xi_{mn}-\\
&(\mathbf{x}_{mn,m}\circ\delta\mathbf{u}_{sn,s}+\mathbf{x}_{sn,s}\circ\delta\mathbf{u}_{mn,m})\Delta\xi_{sn}-\\
&\delta\xi_{sn}\,(\mathbf{x}_{mn,m}\circ\Delta\mathbf{u}_{sn,s}+\mathbf{x}_{sn,s}\circ\Delta\mathbf{u}_{mn,m}),
\end{aligned}\tag{A2.8$_1$}$$

$$
\begin{aligned}
R_s = {} & p_1\,\delta\xi_{sn}\,\Delta\xi_{sn} + p_2\left(\delta\xi_{sn}\,\Delta\xi_{mn} + \delta\xi_{mn}\,\Delta\xi_{sn}\right) + p_3\,\delta\xi_{mn}\,\Delta\xi_{mn} \\
& + \mathbf{x}_{ms} \circ \Delta\delta\mathbf{u}_{sn,s} + \delta\mathbf{u}_{sn,s} \circ \left(\Delta\mathbf{u}_{mn} - \Delta\mathbf{u}_{sn}\right) + \\
& + \left(\delta\mathbf{u}_{mn} - \delta\mathbf{u}_{sn}\right) \circ \Delta\mathbf{u}_{sn,s} + \mathbf{x}_{sn,s} \circ \left(\Delta\delta\mathbf{u}_{mn} - \Delta\delta\mathbf{u}_{sn}\right) + \\
& + \delta\xi_{sn}\left[\mathbf{x}_{ms} \circ \Delta\mathbf{u}_{sn,ss} + \mathbf{x}_{sn,ss} \circ \left(\Delta\mathbf{u}_{mn} - \Delta\mathbf{u}_{sn}\right) - 2\mathbf{x}_{sn,s} \circ \Delta\mathbf{u}_{sn,s}\right] + \\
& + \left[\mathbf{x}_{ms} \circ \delta\mathbf{u}_{sn,ss} + \mathbf{x}_{sn,ss} \circ \left(\delta\mathbf{u}_{mn} - \delta\mathbf{u}_{sn}\right) - 2\mathbf{x}_{sn,s} \circ \delta\mathbf{u}_{sn,s}\right]\Delta\xi_{sn} + \\
& + \left(\mathbf{x}_{mn,m} \circ \Delta\mathbf{u}_{sn,s} + \mathbf{x}_{sn,s} \circ \Delta\mathbf{u}_{mn,m}\right)\Delta\xi_{mn} + \\
& + \delta\xi_{mn}\left(\mathbf{x}_{mn,m} \circ \Delta\mathbf{u}_{sn,s} + \mathbf{x}_{sn,s} \circ \Delta\mathbf{u}_{mn,m}\right).
\end{aligned}
\tag{A2.8$_2$}
$$

After the finite element discretisation R_m and R_s can be put in the matrix form

$$
\begin{aligned}
R_m &= \left[\delta\mathbf{u}_M{}^{\mathbf{T}}, \delta\mathbf{u}_S{}^{\mathbf{T}}\right] \mathbf{R}_m \begin{bmatrix} \Delta\mathbf{u}_M \\ \Delta\delta\mathbf{u}_S \end{bmatrix}, \\
R_s &= \left[\delta\mathbf{u}_M{}^{\mathbf{T}}, \delta\mathbf{u}_S{}^{\mathbf{T}}\right] \mathbf{R}_s \begin{bmatrix} \Delta\mathbf{u}_M \\ \Delta\delta\mathbf{u}_S \end{bmatrix}.
\end{aligned}
\tag{A2.9}
$$

Substitution of (A2.9) and the notation from (A2.4) to Eq. (A2.7) leads to the following relations

$$
\begin{aligned}
\Delta\delta\xi_{mn} &= \left[\delta\mathbf{u}_M{}^T, \delta\mathbf{u}_S{}^T\right]\left(a_{mm}\mathbf{R}_m + a_{ms}\mathbf{R}_s\right)\begin{bmatrix} \Delta\mathbf{u}_M \\ \Delta\mathbf{u}_S \end{bmatrix}, \\
\Delta\delta\xi_{sn} &= \left[\delta\mathbf{u}_M{}^T, \delta\mathbf{u}_S{}^T\right]\left(a_{sm}\mathbf{R}_m + a_{ss}\mathbf{R}_s\right)\begin{bmatrix} \Delta\mathbf{u}_M \\ \Delta\mathbf{u}_S \end{bmatrix}.
\end{aligned}
\tag{A2.10}
$$

In order to define the matrices $\mathbf{R}_m$ and $\mathbf{R}_s$ expressions for the second derivatives of displacements with respect to the local co-ordinates

$$
\begin{aligned}
\delta\mathbf{u}_{mn,mm} &= \mathbf{M}_{mn}\,\delta\mathbf{u}_M, \\
\delta\mathbf{u}_{sn,ss} &= \mathbf{M}_{sn}\,\delta\mathbf{u}_S
\end{aligned}
\tag{A2.11}
$$

and for the linearisation of variations of displacements derivatives with respect to the local co-ordinates

$$
\Delta\delta\mathbf{u}_{mn,m} = \begin{bmatrix} \delta\mathbf{u}_M{}^T\,\mathbf{H}_{d1mn}\,\Delta\mathbf{u}_M \\ \delta\mathbf{u}_M{}^T\,\mathbf{H}_{d2mn}\,\Delta\mathbf{u}_M \\ \delta\mathbf{u}_M{}^T\,\mathbf{H}_{d3mn}\,\Delta\mathbf{u}_M \end{bmatrix},
\tag{A2.12$_1$}
$$

$$\Delta\delta\mathbf{u}_{sn,s} = \begin{bmatrix} \delta\mathbf{u}_S^T \, \mathbf{H}_{d1sn} \, \Delta\mathbf{u}_S \\ \delta\mathbf{u}_S^T \, \mathbf{H}_{d2sn} \, \Delta\mathbf{u}_S \\ \delta\mathbf{u}_S^T \, \mathbf{H}_{d3sn} \, \Delta\mathbf{u}_S \end{bmatrix} \tag{A2.12$_2$}$$

are also required.

The matrices $\mathbf{M}_{mn}$ and $\mathbf{M}_{sn}$ in Eqs. (A2.11) can be calculated explicitly as the derivatives of the subsequent components of the matrices $\mathbf{H}_{mn}$ and $\mathbf{H}_{sn}$ with respect to the local co-ordinates ξ_m and ξ_s. However, due to a very complex form of the displacement approximation for the points on the edges of beams with rectangular cross-sections it is more convenient to use a simplified method employing the finite difference method and small perturbations of the local co-ordinates.

On the other hand, the matrices $\mathbf{H}_{djmn}$ and $\mathbf{H}_{djsn}$ in Eqs. (A2.12) can be obtained by differentiation of the matrices $\mathbf{H}_{mn}$ and $\mathbf{H}_{sn}$ in the same way as the matrices $\mathbf{G}_{djmn}$ and $\mathbf{G}_{djsn}$ were derived from Eqs. (2.54). Hence, one can write down

$$\begin{aligned} \mathbf{H}_{djmn} &= \frac{\partial \mathbf{H}_{jmn}}{\partial \mathbf{u}_M}, \\ \mathbf{H}_{djsn} &= \frac{\partial \mathbf{H}_{jsn}}{\partial \mathbf{u}_S}. \end{aligned} \tag{A2.13}$$

Additionally three auxiliary matrices

$$\begin{aligned} \mathbf{Y}_{11} &= \begin{bmatrix} -\mathbf{H}_{mn}^T \, \mathbf{x}_{mn,m} \\ -\mathbf{H}_{sn}^T \, \mathbf{x}_{sn,n} \end{bmatrix}, \\ \mathbf{Y}_{12} &= \begin{bmatrix} \mathbf{M}_{mn}^T \, \mathbf{x}_{ms} + 2\mathbf{H}_{mn}^T \, \mathbf{x}_{mn,m} + \mathbf{G}_{mn}^T \, \mathbf{x}_{mn,mm} \\ -\mathbf{G}_{sn}^T \, \mathbf{x}_{mn,mm} \end{bmatrix}, \\ \mathbf{Y}_{21} &= \begin{bmatrix} \mathbf{G}_{mn}^T \, \mathbf{x}_{sn,ss} \\ \mathbf{M}_{sn}^T \, \mathbf{x}_{ms} - 2\mathbf{H}_{sn}^T \, \mathbf{x}_{sn,s} - \mathbf{G}_{sn}^T \, \mathbf{x}_{sn,ss} \end{bmatrix} \end{aligned} \tag{A2.14}$$

and a representation of vectors by means of their components

$$\begin{aligned} \mathbf{x}_{ms} &= \left[x_{ms1}, x_{ms2}, x_{ms3}\right]^T, \\ \mathbf{x}_{mn,m} &= \left[x_{mn,m1}, x_{mn,m2}, x_{mn,m3}\right]^T, \\ \mathbf{x}_{sn,s} &= \left[x_{sn,s1}, x_{sn,s2}, x_{sn,s3}\right]^T \end{aligned}$$

are introduced. Finally the matrices $\mathbf{R}_m$ and $\mathbf{R}_s$ can be expressed as

$$\mathbf{R}_m = p_2 \cdot \mathbf{F}_s{}^T \mathbf{F}_s + p_3 \cdot \left(\mathbf{F}_m{}^T \mathbf{F}_s + \mathbf{F}_s{}^T \mathbf{F}_m\right) + p_4 \cdot \mathbf{F}_m{}^T \mathbf{F}_m +$$
$$\mathbf{F}_s{}^T \mathbf{Y}_{11}{}^T + \mathbf{F}_m{}^T \mathbf{Y}_{12}{}^T + \mathbf{Y}_{11} \mathbf{F}_s - \mathbf{Y}_{12} \mathbf{F}_m +$$
$$+ \begin{bmatrix} \left(\mathbf{H}_{mn}{}^T \mathbf{G}_{mn} + \mathbf{G}_{mn}{}^T \mathbf{H}_{mn}\right) & -\mathbf{H}_{mn}{}^T \mathbf{G}_{sn} \\ -\mathbf{G}_{sn}{}^T \mathbf{H}_{mn} & \mathbf{0} \end{bmatrix} +$$
$$+ \begin{bmatrix} \left(\sum_{j=1}^{3} x_{mn,mj} \mathbf{G}_{djmn} + \sum_{j=1}^{3} x_{msj} \mathbf{H}_{djmn}\right) & \mathbf{0} \\ \mathbf{0} & -\sum_{j=1}^{3} x_{mn,mj} \mathbf{G}_{djsn} \end{bmatrix},$$
$$\mathbf{R}_s = p_1 \cdot \mathbf{F}_s{}^T \mathbf{F}_s + p_2 \cdot \left(\mathbf{F}_m{}^T \mathbf{F}_s + \mathbf{F}_s{}^T \mathbf{F}_m\right) + p_3 \cdot \mathbf{F}_m{}^T \mathbf{F}_m +$$
$$\mathbf{F}_s{}^T \mathbf{Y}_{21}{}^T - \mathbf{F}_m{}^T \mathbf{Y}_{11}{}^T + \mathbf{Y}_{21} \mathbf{F}_s - \mathbf{Y}_{11} \mathbf{F}_m +$$
$$+ \begin{bmatrix} \mathbf{0} & \mathbf{G}_{mn}{}^T \mathbf{H}_{sn} \\ \mathbf{H}_{sn}{}^T \mathbf{G}_{mn} & \left(-\mathbf{H}_{sn}{}^T \mathbf{G}_{sn} - \mathbf{G}_{sn}{}^T \mathbf{H}_{sn}\right) \end{bmatrix} +$$
$$+ \begin{bmatrix} \sum_{j=1}^{3} x_{sn,sj} \mathbf{G}_{djmn} & \mathbf{0} \\ \mathbf{0} & \left(-\sum_{j=1}^{3} x_{sn,sj} \mathbf{G}_{djsn} + \sum_{j=1}^{3} x_{msj} \mathbf{H}_{djsn}\right) \end{bmatrix} \tag{A2.15}$$

and their substitution to Eq. (A2.10) yields the discretised form of the kinematic variables $\Delta\delta\xi_{mn}$ and $\Delta\delta\xi_{sn}$.

Appendix 3
Matrices G, H and M in Smoothing Procedures

A3.1 Components of Matrix G

The symbolic algebra computer program Maple 7 is used to calculate the components of the matrix **G** present in the formulation of the contact between the beams modelled by smooth 3D curves (Eqs. (2.34) and (2.35) with a description in Section 4.3.1). Below the commands of the Maple 7 worksheet are given, which allow for calculation and creation of an output ready-to-use Fortran computer code, which computes the matrix **G**. The output itself is quite long and complicated, hence it is not given here. The presented worksheet commands are valid for the inscribed curve Hermite smooth contact finite element only. The matrices **G** for the remaining three smooth contact elements can be obtained in a very similar way.

1. Coefficients of three Hermite's polynomials (4.3): a1, b1, c1, d1; a2, b2, c2, d2; a3, b3, c3, d3, expressed by the components of the position vectors of the nodes xij, where index i denotes the node number and j – the number of the co-ordinate. The following expressions result from the matrix relation (4.10).

```
>restart;
>a1:=(1/8-1/4/wl12)*x11+(1/4/wl12-1/4/wl23)*x21+(-
  1/8+1/4/wl23)*x31;
>a2:=(1/8-1/4/wl12)*x12+(1/4/wl12-1/4/wl23)*x22+(-
  1/8+1/4/wl23)*x32;
>a3:=(1/8-1/4/wl12)*x13+(1/4/wl12-1/4/wl23)*x23+(-
  1/8+1/4/wl23)*x33;
>b1:=1/4/wl12*x11+(-1/4/wl12-1/4/wl23)*x21+1/4/wl23*x31;
>b2:=1/4/wl12*x12+(-1/4/wl12-1/4/wl23)*x22+1/4/wl23*x32;
>b3:=1/4/wl12*x13+(-1/4/wl12-1/4/wl23)*x23+1/4/wl23*x33;
>c1:=(-3/8+1/4/wl12)*x11+(-1/4/wl12+1/4/wl23)*x21+(3/8-
  1/4/wl23)*x31;
>c2:=(-3/8+1/4/wl12)*x12+(-1/4/wl12+1/4/wl23)*x22+(3/8-
  1/4/wl23)*x32;
```

```
>c3:=(-3/8+1/4/wl12)*x13+(-1/4/wl12+1/4/wl23)*x23+(3/8-
 1/4/wl23)*x33;
>d1:=(1/4-1/4/wl12)*x11+(1/2+1/4/wl12+1/4/wl23)*x21+(1/4-
 1/4/wl23)*x31;
>d2:=(1/4-1/4/wl12)*x12+(1/2+1/4/wl12+1/4/wl23)*x22+(1/4-
 1/4/wl23)*x32;
>d3:=(1/4-1/4/wl12)*x13+(1/2+1/4/wl12+1/4/wl23)*x23+(1/4-
 1/4/wl23)*x33;
```

2. Straight line lengths (4.7).

```
>wl:=.5*sqrt((x31-x11)^2+(x32-x12)^2+(x33-x13)^2);
>wl12:=2*sqrt((x21-x11)^2+(x22-x12)^2+(x23-x13)^2)/wl;
>wl23:=2*sqrt((x31-x21)^2+(x32-x22)^2+(x33-x23)^2)/wl;
```

3. Components of the matrix **G** calculated as partial derivatives of (4.10) with respect to the subsequent components of the nodal position vectors.

```
>g11:=xsi^3*(diff(a1,x11))+xsi^2*(diff(b1,x11))+xsi*(diff(c1,
 x11))+diff(d1,x11);
>g12:=xsi^3*(diff(a1,x12))+xsi^2*(diff(b1,x12))+xsi*(diff(c1,
 x12))+diff(d1,x12);
>g13:=xsi^3*(diff(a1,x13))+xsi^2*(diff(b1,x13))+xsi*(diff(c1,
 x13))+diff(d1,x13);
>g14:=xsi^3*(diff(a1,x21))+xsi^2*(diff(b1,x21))+xsi*(diff(c1,
 x21))+diff(d1,x21);
>g15:=xsi^3*(diff(a1,x22))+xsi^2*(diff(b1,x22))+xsi*(diff(c1,
 x22))+diff(d1,x22);
>g16:=xsi^3*(diff(a1,x23))+xsi^2*(diff(b1,x23))+xsi*(diff(c1,
 x23))+diff(d1,x23);
>g17:=xsi^3*(diff(a1,x31))+xsi^2*(diff(b1,x31))+xsi*(diff(c1,
 x31))+diff(d1,x31);
>g18:=xsi^3*(diff(a1,x32))+xsi^2*(diff(b1,x32))+xsi*(diff(c1,
 x32))+diff(d1,x32);
>g19:=xsi^3*(diff(a1,x33))+xsi^2*(diff(b1,x33))+xsi*(diff(c1,
 x33))+diff(d1,x33);
>g21:=xsi^3*(diff(a2,x11))+xsi^2*(diff(b2,x11))+xsi*(diff(c2,
 x11))+diff(d2,x11);
>g22:=xsi^3*(diff(a2,x12))+xsi^2*(diff(b2,x12))+xsi*(diff(c2,
 x12))+diff(d2,x12);
>g23:=xsi^3*(diff(a2,x13))+xsi^2*(diff(b2,x13))+xsi*(diff(c2,
 x13))+diff(d2,x13);
>g24:=xsi^3*(diff(a2,x21))+xsi^2*(diff(b2,x21))+xsi*(diff(c2,
 x21))+diff(d2,x21);
```

```
>g25:=xsi^3*(diff(a2,x22))+xsi^2*(diff(b2,x22))+xsi*(diff(c2,
 x22))+diff(d2,x22);
>g26:=xsi^3*(diff(a2,x23))+xsi^2*(diff(b2,x23))+xsi*(diff(c2,
 x23))+diff(d2,x23);
>g27:=xsi^3*(diff(a2,x31))+xsi^2*(diff(b2,x31))+xsi*(diff(c2,
 x31))+diff(d2,x31);
>g28:=xsi^3*(diff(a2,x32))+xsi^2*(diff(b2,x32))+xsi*(diff(c2,
 x32))+diff(d2,x32);
>g29:=xsi^3*(diff(a2,x33))+xsi^2*(diff(b2,x33))+xsi*(diff(c2,
 x33))+diff(d2,x33);
>g31:=xsi^3*(diff(a3,x11))+xsi^2*(diff(b3,x11))+xsi*(diff(c3,
 x11))+diff(d3,x11);
>g32:=xsi^3*(diff(a3,x12))+xsi^2*(diff(b3,x12))+xsi*(diff(c3,
 x12))+diff(d3,x12);
>g33:=xsi^3*(diff(a3,x13))+xsi^2*(diff(b3,x13))+xsi*(diff(c3,
 x13))+diff(d3,x13);
>g34:=xsi^3*(diff(a3,x21))+xsi^2*(diff(b3,x21))+xsi*(diff(c3,
 x21))+diff(d3,x21);
>g35:=xsi^3*(diff(a3,x22))+xsi^2*(diff(b3,x22))+xsi*(diff(c3,
 x22))+diff(d3,x22);
>g36:=xsi^3*(diff(a3,x23))+xsi^2*(diff(b3,x23))+xsi*(diff(c3,
 x23))+diff(d3,x23);
>g37:=xsi^3*(diff(a3,x31))+xsi^2*(diff(b3,x31))+xsi*(diff(c3,
 x31))+diff(d3,x31);
>g38:=xsi^3*(diff(a3,x32))+xsi^2*(diff(b3,x32))+xsi*(diff(c3,
 x32))+diff(d3,x32);
>g39:=xsi^3*(diff(a3,x33))+xsi^2*(diff(b3,x33))+xsi*(diff(c3,
 x33))+diff(d3,x33);
```

4. Creation of the optimised Fortran code with formulae to calculate the components of the matrix **G**.

```
>with(codegen,fortran);
>fortran(g11,optimized,mode=double,precision=double);
>fortran(g12,optimized,mode=double,precision=double);
>fortran(g13,optimized,mode=double,precision=double);
>fortran(g14,optimized,mode=double,precision=double);
>fortran(g15,optimized,mode=double,precision=double);
>fortran(g16,optimized,mode=double,precision=double);
>fortran(g17,optimized,mode=double,precision=double);
>fortran(g18,optimized,mode=double,precision=double);
>fortran(g19,optimized,mode=double,precision=double);
>fortran(g21,optimized,mode=double,precision=double);
>fortran(g22,optimized,mode=double,precision=double);
```

```
>fortran(g23,optimized,mode=double,precision=double);
>fortran(g24,optimized,mode=double,precision=double);
>fortran(g25,optimized,mode=double,precision=double);
>fortran(g26,optimized,mode=double,precision=double);
>fortran(g27,optimized,mode=double,precision=double);
>fortran(g28,optimized,mode=double,precision=double);
>fortran(g29,optimized,mode=double,precision=double);
>fortran(g31,optimized,mode=double,precision=double);
>fortran(g32,optimized,mode=double,precision=double);
>fortran(g33,optimized,mode=double,precision=double);
>fortran(g34,optimized,mode=double,precision=double);
>fortran(g35,optimized,mode=double,precision=double);
>fortran(g36,optimized,mode=double,precision=double);
>fortran(g37,optimized,mode=double,precision=double);
>fortran(g38,optimized,mode=double,precision=double);
>fortran(g39,optimized,mode=double,precision=double);
```

A3.2 Components of Matrix H

The symbolic algebra computer program Maple 7 is used to calculate the components of the matrix **H** present in the formulation of the contact between the beams modelled by smooth 3D curves (Eqs. (2.36) and (2.37) with a description in Section 4.3.1). Below the commands of the Maple 7 worksheet are given, which allow for calculation and creation of an output ready-to-use Fortran computer code, which computes the matrix **H**. The output itself is quite long and complicated, hence it is not given here. The presented worksheet commands are valid for the inscribed curve Hermite smooth contact finite element only. The matrices **H** for the remaining three smooth contact elements can be obtained in a very similar way.

1. Coefficients of three Hermite's polynomials – the same as for the matrix **G** (Section A3.1)

2. Straight line lengths – the same as for the matrix **G** (Section A3.1)

3. Components of the matrix **H** calculated as partial derivatives of $(4.11)_1$ with respect to the subsequent components of the nodal position vectors.

```
>h11:=3*xsi^2*(diff(a1,x11))+2*xsi*(diff(b1,x11))+(diff(c1,x1
  1));
>h12:=3*xsi^2*(diff(a1,x12))+2*xsi*(diff(b1,x12))+(diff(c1,x1
  2));
>h13:=3*xsi^2*(diff(a1,x13))+2*xsi*(diff(b1,x13))+(diff(c1,x1
  3));
```

```
>h14:=3*xsi^2*(diff(a1,x21))+2*xsi*(diff(b1,x21))+(diff(c1,x2
 1));
>h15:=3*xsi^2*(diff(a1,x22))+2*xsi*(diff(b1,x22))+(diff(c1,x2
 2));
>h16:=3*xsi^2*(diff(a1,x23))+2*xsi*(diff(b1,x23))+(diff(c1,x2
 3));
>h17:=3*xsi^2*(diff(a1,x31))+2*xsi*(diff(b1,x31))+(diff(c1,x3
 1));
>h18:=3*xsi^2*(diff(a1,x32))+2*xsi*(diff(b1,x32))+(diff(c1,x3
 2));
>h19:=3*xsi^2*(diff(a1,x33))+2*xsi*(diff(b1,x33))+(diff(c1,x3
 3));
>h21:=3*xsi^2*(diff(a2,x11))+2*xsi*(diff(b2,x11))+(diff(c2,x1
 1));
>h22:=3*xsi^2*(diff(a2,x12))+2*xsi*(diff(b2,x12))+(diff(c2,x1
 2));
>h23:=3*xsi^2*(diff(a2,x13))+2*xsi*(diff(b2,x13))+(diff(c2,x1
 3));
>h24:=3*xsi^2*(diff(a2,x21))+2*xsi*(diff(b2,x21))+(diff(c2,x2
 1));
>h25:=3*xsi^2*(diff(a2,x22))+2*xsi*(diff(b2,x22))+(diff(c2,x2
 2));
>h26:=3*xsi^2*(diff(a2,x23))+2*xsi*(diff(b2,x23))+(diff(c2,x2
 3));
>h27:=3*xsi^2*(diff(a2,x31))+2*xsi*(diff(b2,x31))+(diff(c2,x3
 1));
>h28:=3*xsi^2*(diff(a2,x32))+2*xsi*(diff(b2,x32))+(diff(c2,x3
 2));
>h29:=3*xsi^2*(diff(a2,x33))+2*xsi*(diff(b2,x33))+(diff(c2,x3
 3));
>h31:=3*xsi^2*(diff(a3,x11))+2*xsi*(diff(b3,x11))+(diff(c3,x1
 1));
>h32:=3*xsi^2*(diff(a3,x12))+2*xsi*(diff(b3,x12))+(diff(c3,x1
 2));
>h33:=3*xsi^2*(diff(a3,x13))+2*xsi*(diff(b3,x13))+(diff(c3,x1
 3));
>h34:=3*xsi^2*(diff(a3,x21))+2*xsi*(diff(b3,x21))+(diff(c3,x2
 1));
>h35:=3*xsi^2*(diff(a3,x22))+2*xsi*(diff(b3,x22))+(diff(c3,x2
 2));
>h36:=3*xsi^2*(diff(a3,x23))+2*xsi*(diff(b3,x23))+(diff(c3,x2
 3));
>h37:=3*xsi^2*(diff(a3,x31))+2*xsi*(diff(b3,x31))+(diff(c3,x3
 1));
```

```
>h38:=3*xsi^2*(diff(a3,x32))+2*xsi*(diff(b3,x32))+(diff(c3,x3
 2));
>h39:=3*xsi^2*(diff(a3,x33))+2*xsi*(diff(b3,x33))+(diff(c3,x3
 3));
```

4. Creation of the optimised Fortran code with formulae to calculate the components of the matrix **H** – the same as for matrix **G** (Section A3.1) with the substitution `hij` instead of `gij`.

A3.3 Components of Matrix M

The symbolic algebra computer program Maple 7 is used to calculate the components of the matrix **M** present in the formulation of the contact between the beams modelled by smooth 3D curves (Eqs. (A2.11) with a description in Section 4.3.1). Below the commands of the Maple 7 worksheet are given, which allow for calculation and creation of output ready-to-use Fortran computer code, which computes the matrix **M**. The output itself is quite long and complicated, hence it is not given here. The presented worksheet commands are valid for the inscribed curve Hermite smooth contact finite element only. The matrices **M** for the remaining three smooth contact elements can be obtained in a very similar way.

1. Coefficients of three Hermite's polynomials – the same as for the matrix **G** (Section A3.1)

2. Straight line lengths – the same as for the matrix **G** (Section A3.1)

3. Components of the matrix **M** calculated as partial derivatives of $(4.11)_2$ with respect to the subsequent components of the nodal position vectors.

```
>m11:=6*xsi*(diff(a1,x11))+2*(diff(b1,x11));
>m12:=6*xsi*(diff(a1,x12))+2*(diff(b1,x12));
>m13:=6*xsi*(diff(a1,x13))+2*(diff(b1,x13));
>m14:=6*xsi*(diff(a1,x21))+2*(diff(b1,x21));
>m15:=6*xsi*(diff(a1,x22))+2*(diff(b1,x22));
>m16:=6*xsi*(diff(a1,x23))+2*(diff(b1,x23));
>m17:=6*xsi*(diff(a1,x31))+2*(diff(b1,x31));
>m18:=6*xsi*(diff(a1,x32))+2*(diff(b1,x32));
>m19:=6*xsi*(diff(a1,x33))+2*(diff(b1,x33));
>m21:=6*xsi*(diff(a2,x11))+2*(diff(b2,x11));
>m22:=6*xsi*(diff(a2,x12))+2*(diff(b2,x12));
>m23:=6*xsi*(diff(a2,x13))+2*(diff(b2,x13));
>m24:=6*xsi*(diff(a2,x21))+2*(diff(b2,x21));
>m25:=6*xsi*(diff(a2,x22))+2*(diff(b2,x22));
```

```
>m26:=6*xsi*(diff(a2,x23))+2*(diff(b2,x23));
>m27:=6*xsi*(diff(a2,x31))+2*(diff(b2,x31));
>m28:=6*xsi*(diff(a2,x32))+2*(diff(b2,x32));
>m29:=6*xsi*(diff(a2,x33))+2*(diff(b2,x33));
>m31:=6*xsi*(diff(a3,x11))+2*(diff(b3,x11));
>m32:=6*xsi*(diff(a3,x12))+2*(diff(b3,x12));
>m33:=6*xsi*(diff(a3,x13))+2*(diff(b3,x13));
>m34:=6*xsi*(diff(a3,x21))+2*(diff(b3,x21));
>m35:=6*xsi*(diff(a3,x22))+2*(diff(b3,x22));
>m36:=6*xsi*(diff(a3,x23))+2*(diff(b3,x23));
>m37:=6*xsi*(diff(a3,x31))+2*(diff(b3,x31));
>m38:=6*xsi*(diff(a3,x32))+2*(diff(b3,x32));
>m39:=6*xsi*(diff(a3,x33))+2*(diff(b3,x33));
```

4. Creation of the optimised Fortran code with formulae to calculate the components of the matrix **M** – the same as for the matrix **G** (Section A3.1) with the substitution `mij` instead of `gij`.

Bibliography

Agelet de Saracibar, C.: A new frictional time integration algorithm for large slip multi-body frictional contact problems. Comput. Meth. Appl. Mech. Eng. 142, 303–334 (1997)

Agelet de Saracibar, C.: Numerical analysis of coupled thermomechanical frictional contact problems, computational model and applications. Arch. Comput. Meth. Eng. 5, 243–302 (1998)

Alart, P., Curnier, A.: A mixed formulation for frictional contact problems prone to Newton like solution methods. Comput. Meth. Appl. Mech. Eng. 92, 353–375 (1991)

Anand, L.: A constitutive model for interface friction. Comput. Mech. 12, 197–213 (1993)

Bathe, K.-J.: Finite Element Procedures. Prentice Hall, Upper Saddle River (1996)

Belytschko, T., Neal, M.O.: Contact-impact by the pinball algorithm with penalty and Lagrange methods. Int. J. Num. Meth. Eng. 10, 579–596 (1991)

Belytschko, T., Daniel, W.J.T., Ventura, G.: A monolithic smoothing-gap algorithm for contact-impact based on the signed distance function. Int. J. Num. Meth. Eng. 55, 101–125 (2002)

Björkman, G., Klarbring, A., Sjödin, B.: Sequential quadratic programming for non-linear elastic contact problems. Int. J. Num. Meth. Eng. 38, 137–165 (1995)

Bruneel, H.C.J., De Rycke, I.: QuickTrace: a fast algorithm to detect contact. Int. J. Num. Meth. Eng. 54, 299–316 (2002)

Bonnet, J., Peraire, J.: An Alternating Digital Tree (ADT) algorithm for 3D geometric searching and intersection problems. Int. J. Num. Meth. Eng. 31, 1–17 (1991)

Boso, D.P., Zavarise, G., Schrefler, B.A.: A formulation for electrostatic-mechanical contact and its numerical solution. Int. J. Num. Meth. Eng. 64, 382–400 (2005a)

Boso, D.P., Litewka, P., Schrefler, B.A., Wriggers, P.: A 3D beam-to-beam contact finite element for coupled electric-mechanical fields. Int. J. Num. Meth. Eng. 64, 1800–1815 (2005b)

Boso, D.P., Litewka, P., Schrefler, B.A., Wriggers, P.: Thermo-electro-mechanical coupling in beam-to-beam contact. In: Soares, M., et al. (eds.) III European Conference on Comput. Mech. Solids Structures and Coupled Problems in Engineering, Lisbon, Portugal, June 5-8 (2006)

Bowden, F.P., Tabor, D.: The Friction and Lubrication of Solids, Part II. Clarendon Press, Oxford (1964)

Chan, S.H., Tuba, I.S.: A finite element method for contact problems in solid bodies. Int. J. Mech. Scien. 13, 615–639 (1971)

Cooper, M.G., Mikic, B.B., Yovanovic, M.M.: Thermal contact conductance. Int. J. Heat Mass Transf. 12, 279–300 (1969)

Coulomb, C.A.: Théorie des machines simples, en ayant égard au frottement de leurs parties et à la roideur des cordages (1779)

Courtney-Pratt, J.S., Eisner, E.: The effect of tangential force on the contact of metallic bodies. Proceedings of the Royal Society of London 238-A, 529–550 (1957)

Crisfield, M.A.: A consistent co-rotational formulation for non-linear, three-dimensional beam-elements. Comput. Meth. Appl. Mech. Eng. 81, 131–150 (1990)

Curnier, A., Alart, P.: A generalized Newton method for contact problems with friction. Journal de Mecanique Theorique et Appliquee 7, 67–82 (1988)

El Abbasi, N., Meguid, S.A., Czekanski, A.: On the modelling of smooth contact surfaces using cubic splines. Int. J. Num. Meth. Eng. 50, 953–967 (2001)

Euler, L.: Sur le frottement des corps solides. Memoirs of Academy of Sciences Berlin 4, 122–132 (1748a)

Euler, L.: Sur la diminution de la resistance du frottement. Memoirs of Academy of Sciences Berlin 4, 133–148 (1748b)

Farin, G.: Curves and Surfaces for Computer Aided Geometric Design. Academic Press Inc., Harcourt Brace Jovanovich Publishers, London, New York (1993)

Feng, Y.T., Owen, D.R.J.: An augmented spatial digital tree algorithm for contact detection in computational mechanics. Int. J. Num. Meth. Eng. 55, 159–176 (2002)

Gawęcki, A.: Elasto-plastic slackened bar structures. Wydawnictwo Politechniki Poznańskiej, Seria Rozprawy Nr. 185 (1987) (in Polish)

Gawęcki, A., Janińska, B.: Computer analysis of slackened systems under variable loads. J. Comput. Appl. Math. 63, 325–332 (1995)

Gawęcki, A., Kuczma, M.S.: Elastic-plastic unilateral contact problems for slackened systems. J. Comput. Appl. Math. 63, 313–323 (1995)

Gawęcki, A., Kuczma, M.S., Krüger, P.: Analysis of slackened skeletal structures by mathematical programming. Comput. Struct. 69, 639–654 (1998)

Glowinski, R., Le Tallec, P.: Finite element analysis in nonlinear incompressible elasticity. In: Finite Element. Special Problems in Solid Mechanics, vol. V. Prentice Hall, Englewood Cliffs (1984)

Greenwood, J.A., Williamson, J.B.P.: The contact of normally-flat surfaces. Proceedings of the Royal Society of London 295-A, 300–379 (1966)

Gun, H.: Boundary element analysis of 3D elasto-plastic contact problems with friction. Comput. Struct. 82, 555–566 (2004)

Hallquist, J.O.: Nike2D: An implicit, finite-deformation, finite element code for analysing the static and dynamic response of two-dimensional solids. Technical Report UCRL-52678, University of California, Lawrence Livermore National Laboratory (1979)

Hertz, H.: Über die Berührung fester elastischer Körper. Journal für die Reine und Angewandte Mathematik 29, 156–171 (1882)

Holm, R.: Electric Contacts: Theory and Applications. Springer, Heidelberg (1981)

Johnson, K.L.: Contact Mechanics. Cambridge University Press, Cambridge (1985)

Jones, R.E., Papadopoulos, P.: A novel three-dimensional contact finite element based on smooth pressure interpolations. Int. J. Num. Meth. Eng. 51, 791–811 (2001)

Kikuchi, N., Oden, J.T.: Contact Problems in Elasticity: A Study of Variational Inequalities and Finite Element Methods. SIAM, Philadelphia (1988)

Kim, J.O., Kwak, B.M.: Dynamic analysis of two-dimensional frictional contact by linear complementarity problem formulation. Int. J. Solids Struct. 33, 4605–4624 (1996)

Klarbring, A.: A mathematical programming approach to three-dimensional contact problems with friction. Comput. Meth. Appl. Mech. Eng. 58, 175–200 (1986)

Klarbring, A.: On discrete and discretized nonlinear elastic structures in unilateral contact (stability, uniqueness and variational principles). Int. J. Solids Struct. 24, 459–479 (1988)

Klarbring, A.: Examples of non-uniqueness and non-existence of solutions to quasistatic contact problems with friction. Ingenieur Archiv. 60, 529–541 (1990)

Klarbring, A., Björkman, G.: A mathematical programming approach to contact problems with friction and varying contact surfaces. Comput. Struct. 30, 1185–1198 (1988)

Konyukhov, A., Schweizerhof, K.: Covariant description for contact between arbitrary curves: general approach for beams, cables and surface edges. In: 8th World Congress on Computational Mechanics/5th European congress on Computational methods in Applied Sciences and Engineering Methods in Mechanics, Venice, Italy, June 30-July 4, CD-ROM, paper a1121 (2008)

Konyukhov, A., Schweizerhof, K.: On a geometrically exact theory for contact interactions. In: 1st International Conference on Computational Contact Mechanics, ICCCM 2009, Lecce Italy, September 16-18 (2009); Abstracts: 81-82

Kragelsky, I.V., Dobychin, M.N., Kombalov, V.S.: Friction and Wear – Calculation Methods. Pergamon Press, Oxford (1982)

Krstulović-Opara, L.: A C1-continuous formulation for finite deformation contact, Dr-Ing Dissertation. Institut für Baumechanik und Numerische Mechanik, Universität Hannover (2001)

Krstulović-Opara, L., Wriggers, P., Korelc, J.: A C^1-continuous formulation for 3D finite deformation frictional contact. Comput. Mech. 29, 27–42 (2002)

Laursen, T.A.: Computational Contact and Impact Mechanics. Springer, Heidelberg (2002)

Laursen, T.A., Simo, J.C.: Algorithmic symmetrization of Coulomb frictional problems using augmented Lagrangians. Comput. Meth. Appl. Mech. Eng. 108, 133–146 (1993a)

Laursen, T.A., Simo, J.C.: A continuum-based finite element formulation for the implicit solution of multibody, large deformation frictional contact problems. Int. J. Num. Meth. Eng. 36, 3451–3485 (1993b)

Li, J., Berger, E.J.: A semi-analytical approach to three-dimensional normal contact problems with friction. Comput. Mech. 30, 310–322 (2003)

Li, S., Qian, D., Liu, W.-K., Belytschko, T.: A meshfree contact detection algorithm. Comput. Meth. Appl. Mech. Eng. 190, 3271–3292 (2001)

Litewka, P.: The penalty and Lagrange multiplier methods in the frictional 3D beam-to-beam contact problem. Civ. Environ. Eng. Reports 1, 189–207 (2005)

Litewka, P.: Hermite polynomial smoothing in beam-to-beam frictional contact problem. Comput. Mech. 40(6), 815–826 (2007)

Litewka, P.: A 3D curve smoothing method preserving nodes for the beam-to-beam contact. In: 8th World Congress on Computational Mechanics/5th European congress on Computational methods in Applied Sciences and Engineering Methods in Mechanics, Venice, Italy, June 30 - July 4, CD-ROM, paper a2916 (2008)

Litewka, P., Wriggers, P.: Contact between 3D beams with rectangular cross-sections. Int. J. Num. Meth. Eng. 53, 2019–2041 (2002a)

Litewka, P., Wriggers, P.: Frictional contact between 3D beams. Comput. Mech. 28, 26–39 (2002b)

Litewka, P., Wriggers, P.: Frictional beam-to-beam contact by the Lagrange multipliers method. In: Oñate, E., Owen, D.R.J. (eds.) Complas VII 7th International Conference on Computational Plasticity, Barcelona, Spain, April 7-10 (2003)

Litewka, P., Wriggers, P., Rakowski, J.: 3D beam finite element for large displacement analysis. In: ECCM 2001, 2nd European Conference on Computational Mechanics Solids, Structures and Coupled Problems in Engineering, Cracow, Poland, June 26-29, 2001 (2001)

Luenberger, D.G.: Linear and Nonlinear Programming. Addison-Wesley, Reading (1984)

Maker, B.N., Laursen, T.A.: A finite element formulation for the rod/continuum interactions: the-one dimensional sliding. Int. J. Num. Meth. Eng. 37, 1–18 (1994)

Michałowski, R., Mróz, Z.: Associated and non-associated sliding rules in contact friction problems. Arch. Mech. 30, 259–276 (1978)

Moreau, J.J.: On unilateral constraints, friction and plasticity. In: Capriz, G., Stampacchia, G. (eds.) New Variational Techniques in Mathematical Physics, pp. 175–322. CIME, Roma (1974)

Munjiza, A., Andrews, K.R.F.: NBS contact detection algorithm for bodies of similar size. Int. J. Num. Meth. Eng. 43, 131–149 (1998)

Munjiza, A., Owen, D.R.J., Bicanic, N.: A combined finite-discrete element method in transient dynamics of fracturing solids. Eng. Comput. 12, 145–174 (1995)

Oden, J.T., Martins, J.A.C.: Models and computational methods for dynamic friction phenomena. Comput. Meth. Appl. Mech. Eng. 52, 527–634 (1986)

Pinto da Costa, A., Martins, J.A.C.: The evolution and rate problems and the computation of all possible evolutions in quasi-static frictional contact. Comput. Meth. Appl. Mech. Eng. 192, 2791–2821 (2003)

Persson, B.N.J.: Sliding Friction, Physical Principles and Applications. Springer, Heidelberg (2000)

Puso, M.A., Laursen, T.A.: A 3D contact smoothing using Gregory patches. Int. J. Num. Meth. Eng. 54, 1161–1194 (2002)

Puso, M.A., Laursen, T.A.: A mortar segment-to-segment contact method for large deformation solid mechanics. Comput. Meth. Appl. Mech. Eng. 193, 601–629 (2004)

Simo, J.C., Laursen, T.A.: An augmented Lagrangian treatment of contact problems involving friction. Comput. Struct. 42, 97–116 (1992)

Song, S., Yovanovic, M.M.: Explicit relative contact pressure expression: Dependence upon surface roughness parameters and Vickers microhardness coefficients. AIAA Journal 87-0152 (1987)

Stadler, M., Holzapfel, G.A., Korelc, J.: C^n-continuous modelling of smooth contact surfaces using NURBS and application to 2D problems. Int. J. Num. Meth. Eng. 57, 2177–2203 (2003)

Stupkiewicz, S.: Extension of the node-to-segment contact element for surface-expansion-dependent contact laws. Int. J. Num. Meth. Eng. 50, 739–759 (2001)

Stupkiewicz, S., Mróz, Z.: A model of third body abrasive friction and wear in hot metal forming. Wear 231, 124–138 (1999)

Sung, J.H., Kwak, B.M.: Large displacement dynamic analysis with frictional contact by linear complementarity formulation. Comput. Struct. 80, 977–988 (2002)

Wang, S.P., Nakamachi, E.: The inside-outside contact search algorithm for finite element analysis. Int. J. Num. Meth. Eng. 40, 3665–3685 (1997)

Williams, J.R., O'Connor, R.: A linear complexity intersection algorithm for discrete element simulation of arbitrary geometries. Eng. Comput. 12, 185–201 (1995)

Williams, J.R., O'Connor, R.: Discrete element simulation and the contact problem. Arch. Comput. Meth. Eng. 6, 279–304 (1999)

Wilson, E.A., Parsons, B.: Finite element analysis of elastic contact problems using differential displacements. Int. J. Num. Meth. Eng. 2, 387–395 (1970)

Woo, K.L., Thomas, T.R.: Contact of rough surfaces: A review of experimental works. Wear 58, 331–340 (1980)

Wriggers, P.: On consistent tangent matrices for frictional contact problems. In: Pande, G., Middleton, J. (eds.) Proceedings of NUMETA 87. M Nijhoff Publishers, Dordrecht (1987)

Wriggers, P.: Computational Contact Mechanics. John Wiley & Sons Ltd., Chichester (2002)
Wriggers, P., Krstulović-Opara, L.: Development of 2D smooth polynomial frictional contact elements based on symbolic approach. In: Wunderlich, W. (ed.) Proceedings of ECCM 1999 European Conference on Computational Mechanics Solids, Structures and Coupled Problems in Engineering, München, Germany, August 31- September 3 (1999)
Wriggers, P., Miehe, C.: On the treatment of contact constraints within coupled thermomechanical analysis. In: Desdo, B., Stein, E. (eds.) Proceedings of EUROMECH, Finite Inelastic Deformations. Springer, Berlin (1992)
Wriggers, P., Zavarise, G.: On the application of augmented Lagrangian techniques for nonlinear constitutive laws in contact interfaces. Comm. Appl. Num. Meth. 9, 815–824 (1993a)
Wriggers, P., Zavarise, G.: Thermomechanical contact – a rigorous but simple numerical approach. Comput. Mech. 46, 47–53 (1993b)
Wriggers, P., Zavarise, G.: On contact between three-dimensional beams undergoing large deflections. Comm. Num. Meth. Eng. 13, 429–438 (1997)
Wriggers, P., Simo, J., Taylor, R.: Penalty and augmented Lagrangian formulations for contact problems. In: Middleton, J., Pande, G. (eds.) Proceedings of NUMETA Conference, Balkema, Rotterdam (1985)
Wriggers, P., Van, T.V., Stein, E.: Finite-element-formulation of large deformation impact-contact problems with friction. Comput. Struct. 37, 319–333 (1990)
Young, W.C., Budynas, R.: Roark's Formulas for Stress and Strain. McGraw-Hill Professional, New York (2001)
Zavarise, G.: Problemi termomeccanici di contatto – aspetti fisici e computazionali, PhD Thesis. Instituto di Scienza e Tecnica delle Construzioni, Universita de Padova, Padova, Italia (1991)
Zavarise, G., Wriggers, P.: Contact with friction between beams in 3-D space. Int. J. Num. Meth. Eng. 49, 977–1006 (2000)
Zavarise, G., Wriggers, P., Schrefler, B.A., Stein, E.: Real contact mechanisms and finite element formulation – a coupled thermomechanical approach. Int. J. Num. Meth. Eng. 35, 767–786 (1992)
Zavarise, G., Wriggers, P., Schrefler, B.A.: On augmented Lagrangian algorithms for thermomechanical contact problems with friction. Int. J. Num. Meth. Eng. 38, 2929–2949 (1995)
Zavarise, G., Bacchetto, A., Gänser, H.P.: Frictional heating in contact mechanics A methodology to deal with high temperature gradients. Comput. Mech. 35, 418–429 (2005)
Zienkiewicz, O.C.: The Finite Element Method, vol. 1. Butterworth Heinemann, Butterworths (2000)
Zhong, Z.H., Nilsson, L.A.: A contact searching algorithm for general contact problems. Comput. Struct. 33, 197–209 (1989)

GPSR Compliance

The European Union's (EU) General Product Safety Regulation (GPSR) is a set of rules that requires consumer products to be safe and our obligations to ensure this.

If you have any concerns about our products, you can contact us on ProductSafety@springernature.com

In case Publisher is established outside the EU, the EU authorized representative is:

Springer Nature Customer Service Center GmbH
Europaplatz 3
69115 Heidelberg, Germany

Batch number: 10583315

Printed by Printforce, the Netherlands